SpringerBriefs in Aging

More information about this series at http://www.springer.com/series/10048

Belinda Yuen · Špela Močnik ·
Freya C.H. Yu · Winston Yap

Ageing-Friendly Neighbourhoods in Singapore, Asia-Pacific, Europe and North America

An Annotated Bibliography

 Springer

Belinda Yuen
Lee Li Ming Programme in Ageing
Urbanism, Lee Kuan Yew Centre
for Innovative Cities
Singapore University of Technology
and Design
Singapore, Singapore

Špela Močnik
Lee Li Ming Programme in Ageing
Urbanism, Lee Kuan Yew Centre
for Innovative Cities
Singapore University of Technology
and Design
Singapore, Singapore

Freya C.H. Yu
Lee Li Ming Programme in Ageing
Urbanism, Lee Kuan Yew Centre
for Innovative Cities
Singapore University of Technology
and Design
Singapore, Singapore

Winston Yap
Lee Li Ming Programme in Ageing
Urbanism, Lee Kuan Yew Centre
for Innovative Cities
Singapore University of Technology
and Design
Singapore, Singapore

ISSN 2211-3231   ISSN 2211-324X   (electronic)
SpringerBriefs in Aging
ISBN 978-3-030-38287-2   ISBN 978-3-030-38288-9   (eBook)
https://doi.org/10.1007/978-3-030-38288-9

This Springer imprint is published by the registered company Springer Nature Switzerland AG
The registered company address is: Gewerbestrasse 11, 6330 Cham, Switzerland

Foreword

Globally, population ageing is gaining research and policy precedence. For the first time in human history, there are more people over the age of 65 than under 5. As the life space of individuals shrink with age, the task of enabling them to remain autonomous within familiar environments while ameliorating the effects of functional decline has become urgent. The discourse on age-friendly environments has understandably attracted increasing attention.

An annotated bibliography in relation to age-friendly environment and cities is timely and well-positioned to guide this emergent need. It serves as a quick informed guide to the research terrain, which will be helpful for researchers seeking to define a new focus and avoid duplication. It is informative on the rich body of the literature that is accessible.

This annotated bibliography is the third of a collective series on critical topics of ageing. Past annotations covered research on housing, arts and culture for older population. The Lee Kuan Yew Centre for Innovative Cities under its Lee Li Ming Programme in Ageing Urbanism has recently completed a seminal study on Innovative Planning and Design of Age-Friendly Neighbourhoods in Singapore, funded by the National Research Foundation and Ministry of National Development, Singapore. The Centre continues to promote research and innovation on urban issues including Future of Cities, Urban Environmental Sustainability.

Heng Chee Chan
Ambassador-at-Large and Chairman
Lee Kuan Yew Centre for Innovative Cities
Singapore University of Technology and Design
Singapore, Singapore

Acknowledgements We gratefully acknowledge the contribution of SUTD Library towards the compilation of grey literature on Singapore in this annotated bibliography. In particular, we thank the excellent support of University Librarian Judy Teo and her team led by Joel Teo and their colleagues, Jessie Tang and Li Zhen Yan. This volume draws on research undertaken under the Lee Li Ming Programme in Ageing Urbanism at the Lee Kuan Yew Centre for Innovative Cities, Singapore University of Technology and Design. We are deeply grateful to Mrs Lee Li Ming for funding our ageing research.

Contents

Chapter 1
State of Ageing-Friendly City in Singapore

1.1 Introduction

Singapore's population is rapidly ageing due to a combination of low birth rates and increasing longevity. The number of Singaporean residents aged 65 years and older is projected to increase from the present 540,000–900,000 by 2030 (Department of Statistics 2019). To address the changing needs of demographic ageing, Singapore has started to implement ageing-friendly city policies and solutions (Yuen and Soh 2017).

In the residential neighbourhood, Singapore's approach to creating ageing-friendly environments has three interrelated aims: (1) to ensure that older adults remain physically active and healthy; (2) to encourage older adults to be socially active and minimise social isolation; and (3) to support ageing in place for as long as possible. Ageing in place in Singapore is closely related to neighbourhood life and amenities. The emphasis is on age-friendly housing, neighbourhoods, towns and communities as individual lifespace becomes constricted with age (Ageing Planning Office, n.d.). We prefer the term "ageing-friendly" as it presents a more all-encompassing term that holistically addresses older adults' lived experience. This term is derived from lifespan developmental psychology and proposes an integrated model of ageing well. The model comprises five interrelated concepts that reflect older adults' needs: continuity, compensation, connection, contribution and challenge (Scharlach 2012). To attain these goals requires adequate supports like social and physical environments, in particular, the immediate neighbourhoods that provide opportunities for older adults to fulfil their needs. However, as most of the literature uses the term 'age-friendly', we will comply and continue with this term in this publication unless ageing-friendly is specifically mentioned in the documents.

This chapter examines the current state of policy, practice and research on age-friendly neighbourhoods in Singapore. It traces the evolution of age-friendly policies and identifies the practices and emerging research on age-friendly city/neighbourhood. The World Health Organization (WHO) Age-Friendly Cities (AFC) framework is used as the reference and age-friendly policies, practices and research are explored under the 8 age-friendly cities domains: outdoor spaces and buildings; transportation; housing; social participation; respect and social inclusion; civic participation and employment; communication and information; and community support and health services (WHO 2007). Even though Singapore is not a member of the WHO Global Network for Age-friendly Cities and Communities, its age-friendly initiatives are many and deserve closer examination.

1.2 Policy

This section surveys Singapore's age-friendly policies over the decades, from the time of its first age-friendly policy in the 1980s to the present day. Policies are considered age-friendly if they address one or more of the 8 key domains of age-friendliness in the WHO age-friendly cities checklist (WHO 2007).

1.2.1 Rationale for Age-Friendly City Policies

The need for communities and the city to be age-friendly has been recognised as a key issue by Singapore policymakers since the 1980s. The national discourse for age-friendly policies was initially borne out of economic considerations. Singapore's over-reliance on its manufacturing industry made clear its vulnerability to global macro-economic forces during the 1984–1986 recession (Doran and Dixon 1991). Moving forward, the Singapore government laid ambitious plans to diversify the economy into other sectors of banking and financial services, knowledge-intensive industries, air and seaport services, etc. in an effort to promote more sustainable economic growth. A competitive and high calibre workforce was essential to achieve this goal. Unfortunately, previous decades of population growth control had proven decidedly effective in reducing birth rates (Graham 1995). Concomitantly, improvements in standard of living and medical care led to increased life expectancy. The resulting demographic shift precipitated a need to create an age-friendly environment where older Singaporeans are enabled to lead productive and fulfilling lives in their later years.

  Since then, the Singapore government has envisioned, formulated and implemented many policies to ameliorate the effects of an ageing society. For example, recognising from the outset that caring for an ageing population can impose a heavy financial burden, the government designs policies that strive to improve individual financial security and wellbeing. This approach encourages a shared autonomy care model among various societal stakeholders where the individual, family and the community are the primary mode of eldercare and the government's role is to support and enable. Over time, the purview of policy has expanded since its incipient days to include increasing focus on the built environment.

1.2.2 1980s—Age-Friendly Policy Domains (Respect and Social Inclusion; Civic Participation and Employment, Community Support and Health Services)

The Committee on the Problems of the Aged was established in 1982 under the chairmanship of then Minister for Health, Mr. Howe Yoon Chong. The committee was the first high-level inter-ministerial committee appointed to study the implications of an ageing population and recommend solutions to its challenges as the proportion of older adults aged 65 years and above was increasing (crude death rate of 4.9 per 1000 residents in 1980) even though the population was growing (crude birth rate of 17.6 per 1000 residents in 1980) and predominantly young (median age of 24.4 years in 1980). In their report released two years later, a key recommendation was the need to improve societal attitudes towards older people and ageing (Ministry of Health 1984). The report highlighted how older adults were prevalently viewed as sick and disabled instead of being an invaluable asset given their wealth of experience. The traditional values of filial piety towards older people were emphasised as important moral values to inculcate among younger generations.

It further stressed that older adults should be able to maintain financial independence and security as they age. To achieve this, three major changes were proposed: (1) to raise current retirement age from 55 to 60 years; (2) to offer diverse working arrangements such as part-time, flexi-time or work-from-home jobs; and (3) to restructure the disbursement of individual Central Provident Fund savings (retirement funds) to an annuity stream over the current lump sum option in post-retirement old age. The committee further recommended for regular health screening and activities (e.g. exercise) to be integrated into the daily routines of older adults to keep them healthy and active.

These findings were reflected in a later report: Community-Based Programmes for the Aged by the Advisory Council of the Aged in 1988. The report reviewed community-based programmes for older persons and reiterated the need for appropriate and sufficient healthcare facilities at the neighbourhood level. Crucially, it advocated for the inclusion of social participation as part of the policy agenda. This could be achieved through, e.g. the provision of cross-generational activities, volunteerism and continuing education options to enable the older person to remain socially active. Policy recommendations included the development of a 'University of the Third Age' as a platform for older adults' continuing education and the development of a 'national cadre of retiree volunteers'.

1.2.3   1990s—Age-Friendly Policy Domains (Communication and Information; Community Support and Health Services; Transportation)

Building upon the precepts of age-friendly policies established in earlier decades, the predominant consensus of the 1990s was the need to focus on an integrative and holistic approach to ageing. The main inchoation of a coordinated approach came from the Inter-Ministerial Committee on Healthcare for the Elderly in 1997. Involving government ministries from health and community development, its committee members also spanned voluntary welfare organisations and professionals including geriatric specialists. It was perhaps no coincidence that the multi-disciplinary stakeholders gave rise to a set of multilateral solutions and for the first time, urban issues were addressed in their entirety. This was evident from the multi-dimensional solutions in the 1999 report, which jointly addressed community support, healthcare and social issues. The discussion was focused more on community-level solutions rather than the previously provincial focus on individual-level issues. It was in this report that an Eldercare Master Plan (intended for financial years 2001–2005) to equip neighbourhoods with essential social support services was first mentioned.

Concomitantly, another Inter-Ministerial Committee on the Ageing Population (IMC) was set up in 1998. This committee built upon the work of previous national committees to map a set of desired outcomes and a robust policy framework to anticipate future ageing-related challenges. The IMC recommended the idea of 'ageing in place' to enable older residents to continue living in their homes and communities through the creation of an age-friendly urban environment. It was highlighted that physical transportation and the integration of key amenities and services would be critical towards the success of this approach.

Another key recommendation was the need to involve 'many helping hands' (individual, family, community, state partnership) and strengthen community-based step-down care. Since older adults spend an increasing majority of their time within their neighbourhoods, community-based care can potentially be a cost-effective option in rendering local and direct eldercare services. Policies were designed with the aim to improve financial security and enable families to care for seniors. The IMC proposed a range of policies, which included a severe disability insurance scheme, MediFund[1] as well as community befriending services. Aside from ensuring financial solvency and creating age-friendly physical environments, the committee also recognised the social underpinnings for an age-friendly environment. Personal responsibility, deference to older generations and lifelong planning were raised as desired values essential for helping older adults to remain as active and contributing members of their communities.

---

[1]MediFund is an endowment fund set up by the government to help needy Singaporeans with their medical bill payments.

1.2.4 2000s—Age-Friendly Policy Domains (Housing; Outdoor Spaces and Buildings; Social Participation)

On the threshold of the 2000s, a Service Review Committee (SRC) was set up in 1999 to develop a blueprint for age-friendly services. Drawing on the 1998 IMC findings, the SRC completed and presented its recommendations in the form of the Eldercare Master Plan (financial years 2001–2005) in 2000. The Eldercare Master Plan covered three key components.

Component 1: Establishing physical infrastructure and local community service delivery system

A dual-tiered community service delivery system, which targeted both community and neighbourhood services, was proposed. At the community level, Multi-service Centres would be introduced as one-stop community service centres (e.g. daycare, rehabilitative care, home care services, etc.) while Neighbourhood Links would be commissioned as drop-in centres in local neighbourhood. These drop-in centres would provide information as well as activities and connection opportunities for older adults. The purpose was to enable more streamlined delivery of services while providing increased opportunities to mainstream active participation for older adults within communities.

Component 2: Restructuring funding for service providers to give focus on service affordability

As part of the initiative to encourage voluntary welfare organisations to play a more proactive role in the provision of age-friendly facilities and programmes, the government initiated the Government Financial Assistance Scheme. However, the scheme had some operation issues. For example, due to the cost-sharing arrangement with the government, service providers were not incentivised to economise operations because any surplus incurred would simply reduce the amount of government funding. Another issue was the lack of accountability of service affordability for end-users. To address these concerns, a revised funding framework was proposed. The revised framework adopted an open tender process to evaluate the quality and economy of service providers. In addition, a sliding subsidy scale was proposed to provide higher subsidies for lower income groups so as to ensure service affordability.

Component 3: Ensuring a continuum of activities and services for ambulant older adults, people with frailty and their caregivers

An evaluation conducted by the SRC on existing social participation policies (e.g. senior citizens club, retirees' club, seniors' activity centre, mutual help scheme) had found these policies to be highly effective in improving and sustaining health among ambulant older adults. Encouraged by these findings, the SRC planned to allocate more funds to scale up these initiatives for the expanding ageing population. The committee also highlighted the need to integrate existing geriatric service with domiciliary services to provide a seamless transition for step-down care.

As part of its recommendations, geriatric departments in hospitals were tasked to lead the development of step-down care. Other aspects of the eldercare service eco-system were recognised. For example, counselling services were mentioned as a vital element to help ensure that programmes would meet the needs of older adults. A new Case Management Service would be introduced to help older adults to navigate the entire range of eldercare services. The aim was to provide appropriate guidance for older adults and their families while ensuring the economy of service allocation. Reprieve for caregivers was also recognised as a salient concern in eldercare. The committee emphasised the need to alleviate caregivers' burden through training and information programmes.

Succeeding the SRC plan was the inter-agency Committee on Ageing Issues (CAI). Set up in 2004, the committee comprised of representatives from the government, public, private and academic sectors. Its mandate was to provide holistic strategies to achieve the vision of 'Successful Ageing for Singapore'. Four main recommendations were proposed in its 2006 Report on the Ageing Population (Ministry of Social and Family Development 2006): provide age-friendly housing; create barrier-free environments; supply affordable community-based support services; and promote an active lifestyle among senior citizens.

The committee acknowledged that housing would need to adapt to the changing needs of older people to enable them to age in place. Housing modification subsidies were proposed by the committee to make retrofitting of age-friendly home features financially feasible. This was complemented with a reverse mortgage scheme to help older homeowners to monetise their housing wealth in later years. Barrier-free environments would be implemented to improve mobility and lifespace of older adults. The committee recommended improvements to outdoor public spaces and public transportation, e.g. wheelchair accessibility, easing gradation change for older adults.

On healthcare services, the committee emphasised the importance of accessible healthcare to all and recommended increased subsidies for low-income older Singaporeans. Noting that the participation rate for sports activity was generally low amongst older adults, the committee recommended the establishment of a S$10 million Golden Opportunities Fund to support activities initiated by and organised for older adults. A Ministerial Committee on Ageing was established under the purview of the Ministry of Health in 2007 to implement the recommendations from the CAI report.

1.2.5   2010s—Age-Friendly Policy Domains (Housing; Outdoor Spaces and Buildings; Social Participation)

Since 2010, the national policy agenda had focused centrally on ageing in place and transforming Singapore into an age-friendly city where Singaporeans can age in place confidently and gracefully. Following a successful pilot in Marine Parade neighbourhood, a City for All Ages initiative was introduced in 2011 to create age-friendly communities across Singapore. The approach was to develop a people-centric, ground-up action programme that would create and implement innovative solutions that addressed the needs of older people. It would identify specific precincts as living laboratories to assess the needs of older adults and test out new ideas that could help them age gracefully. A hallmark was its whole-of-government solution process, working across agency lines of responsibility, across public and people sectors, and across government and community settings, so as to achieve new and relevant services that would support the needs of an ageing population more holistically and seamlessly.

As older people's needs differed, various strategies were considered. One key strategic thrust was to ensure that older adults were able to live independently and confidently within their communities. This was achieved through policies that were aimed at creating safe and secure neighbourhoods. Another policy direction was to raise older people's awareness of good health through regular health screening and information sharing for dietary and lifestyle guidance so as to help them better manage their own health and seek early medical attention when needed. The City for All Ages projects also stressed the importance of social capital and the development of a conducive environment that promoted social interaction so as to reduce the risk of social isolation, which was a rising problem among older Singaporeans. The number of socially isolated older adults was projected to triple, from the present 35,000 to 83,000 by 2030.

More recently, in 2016, the Ministerial Committee on Ageing released an Action Plan for Successful Ageing (Ministerial Committee on Ageing 2016). The committee engaged in extensive public consultation (over 4000 Singaporeans from all walks of life, 50 focus group discussions, online consultations, listening points at public spaces) and discussed 10 topics including employability, lifelong learning, senior volunteering, health and wellness, social engagement, aged care services, housing, transport, public spaces and research on ageing. The S$3 billion national plan included more than 70 initiatives in 12 areas, covering health and wellness; learning, volunteerism; employment; housing; transport; public spaces; respect and social inclusion; retirement adequacy; healthcare and aged care; protection for vulnerable; seniors and research. The vision was to realise successful ageing in 3 main areas:

Area 1: Opportunities for all ages

Recognising that longevity is an opportunity, Singapore would become a place where all including older adults can continually learn, grow and achieve their fullest potential. The plan highlighted that a key part of providing opportunities for ageing was to ensure that older adults were able to live meaningful and active lives. Several issues were considered, from retirement adequacy to employment, lifelong learning, health, social participation and volunteerism. As part of the policy to improve work opportunities for older adults, the employment age would be raised from 65 to 67 years by 2017 and a new workplace health programme would target 120,000 workers aged 40 years and older. A National Silver Academy (to train an estimated 30,000 older adults) would offer older adults the opportunity to pursue their learning interests within an intergenerational environment, thus helping to foster intergenerational understanding while providing lifelong learning opportunities.

The plan recognised that to fully capitalise on the opportunities for active ageing, older adults must be in good health. The Ministry of Health would launch a new National Senior's Health Programme. Expected to affect some 400,000 older adults, the programme would have an educational component where older adults would be proffered information on dementia, fall prevention, dietary, medicine and other key health issues related to ageing. The programme included a mobile application service, which enabled older adults to track their medical records and vaccination history. A community dental programme would also be piloted as part of the effort to bring preventive health services to the seniors.

Another major initiative was to improve civic and social participation through volunteerism. Volunteerism was acknowledged as an effective way for older adults to stay active and engaged in their communities. The plan was to champion a national movement for senior volunteerism and recruit 50,000 senior volunteers by 2030. A diverse range of volunteer activities was emphasised as the way to meet the varied interests of older adults and attract new volunteers. Adequate support for volunteers was equally important to the process and could take various forms, e.g. providing government support for out-of-pocket travel/meal expenses, conducting training and information sessions to help volunteers function more effectively, recognising volunteers for their contributions to community.

Area 2: Kampung for all ages

Emphasising inter-generational harmony, Singapore would become a caring and inclusive society where older adults could age actively and happily. Importantly, they would be respected and embraced as an integral part of a cohesive community. The plan emphasised that a crucial component of enabling ageing in place was to ensure that the neighbourhood environment was inclusive and friendly towards older adults. Action areas covered hardware, e.g. design of age-friendly spaces, and software issues, e.g. co-location of eldercare and childcare services to promote intergenerational bonding and understanding from a young age. Through programming of intergenerational activities, opportunities would be presented for older adults to share their experiences with younger generations while allowing younger generations to understand and become more empathetic towards the older person's challenges. To honour their achievements, a Passion Silver Concession Card was launched in 2016 with the support of the People's Association and Singapore Business Federation for older adults aged 60 years and older, giving them benefits and privileges at various services, shops and public transportation. They could also use the concession card to activate longer crossing time at Green Man-Plus pedestrian crossings to lengthen the crossing time.

Area 3: City for all ages

Reiterating the aspiration to transform Singapore into an age-friendly city where Singaporeans could live well and age confidently in place, the goal was for Singapore to become recognised globally for its economic success as well as a model for successful ageing. The plan identified age-friendly housing and mobility as the main strategies for achieving an age-friendly city for all ages. Future housing options would support older people's changing needs (e.g. safer, more senior-friendly) while transport improvement would re-define their travel experience and make it easier, safer and more comfortable for older adults to make their own way around Singapore.

A huge emphasis of this initiative was to create neighbourhoods that were more walkable to encourage more physical and social activity amongst older adults. Action areas would cover infrastructure improvements such as the implementation of 35 Silver Zones by 2020 to reduce vehicular traffic speed in areas with high concentration of older pedestrians, installation of Green Man-Plus at 1000 traffic junctions to extend pedestrian crossing time for older adults, provision of 41 lifts at overhead pedestrian bridges by 2018. To encourage older adults to walk further, seats would be provided at 50 metres interval so that older adults could rest along their journey. Bus and mass rapid transit (MRT) transport would be accordingly improved for the comfort and travel needs of older adults, e.g. to improve connectivity by having more bus routes through mature housing estates, to enhance ridership experience by improving accessibility to MRT stations through the installation of lift services, improving wayfinding through clear and legible wayfinding signs.

Social participation was another key domain for achieving an age-friendly city for all ages. To actively engage both ambulant and frail older adults, Active Ageing Hubs would be integrated into residential neighbourhoods. These hubs would provide a mix of active ageing and social learning activities as well as day-care and rehabilitative care services for older adults. Respite care services would also be planned for caregivers. An estimated 10 active ageing hubs would be developed in new public housing estates by 2020.

1.3 Scanning the Practice

Age-friendly neighbourhood initiatives and good practices are growing in Singapore, especially in recent years, reflecting changing policy development and increased attention on the built environment. Using the WHO Age-friendly Cities framework, the overview on age-friendly practices in Singapore is discussed under 8 issues: outdoor spaces and buildings; transportation; housing; social participation; respect and social inclusion; civic participation and employment; communication and information; and community support and health services. The list of initiatives is illustrative, not exhaustive, as only the government-led initiatives are addressed to offer but a glimpse into the translation of policy into routine practice.

1.3.1 Outdoor Spaces and Buildings

The practice is largely centred on the creation of age-friendly spaces to improve older people's mobility and activity in outdoor spaces.

Silver Zones

First announced in 2014, the Land Transport Authority has introduced Silver Zones in residential neighbourhoods to improve the pedestrian safety of older adults. There are currently 15 silver zones and they have reduced senior pedestrian accident rates by 75%. Silver Zones have several traffic calming features that help reduce car speeding such as chicanes and lanes with reduced width, speed limits and rest points in the middle of the road for older adults to cross the road more easily (Land Transport Authority, n.d.; Song, n.d.). They are generally implemented in areas where there have been senior pedestrian accidents or a high proportion of older residents and their frequently used amenities like clinics, hospitals.

Therapeutic Gardens

The National Parks Board has developed the first outdoor therapeutic garden in 2016. To date, there are four such gardens across Singapore. Therapeutic gardens are outdoor public gardens designed using evidence-based design principles such as attention restoration theory, stress reduction theory and therapeutic horticulture programmes involving nature to cater to the physical, psychological and social needs of park users (National Parks Board 2019). The emphasis is to engage the visitor's senses and stimulate restoration and mental wellbeing.

Senior-friendly Towns

The focus is to improve housing estates, new and old, to make them more age-friendly and responsive to the needs of older residents. This involves universal barrier-free access and micro infrastructure improvement such as providing more rest stops and benches, replacing metal drain covers with less slippery concrete covers, adding colour contrast to more clearly mark level change in outdoor spaces (Ministry of Health, n.d.b). Other local projects implemented under the City for All Ages plan of action to build senior-friendly communities include the Tanglin-Cairnhill bed-side switch programme that installs a two-way light switch by older adults' beds in order to enable them to live independently and confidently. Another project is the Community for Successful Ageing at Whampoa, a partnership project between Tsao Foundation and Whampoa community to introduce primary care mobile clinic, a care management system and community activities for older residents to help them remain healthy, active and part of the community.

1.3.2 Transportation

The effort is towards creating a seamless land transport system that is accessible to all users. It is part of the national action plan for successful ageing that aims to build a city, which enables the older person to lead an active life and age in place confidently.

Green Man-Plus

Green Man-Plus, introduced since October 2009, allows older adults and those with disabilities to tap their senior citizen concession or Green Man-Plus cards at signalised road crossings to have more time to cross the roads. The time can be extended up to 26 s, depending on the width of the road crossing (Land Transport Authority 2013, 2015). Green Man-Plus has been installed in more than 700 pedestrian crossings. As with the Silver Zones, their location is determined by factors such as the density of older population and nearby senior facilities like polyclinic, park.

Senior Citizen Concession Card

Older Singaporeans aged 60 years and older can use a Senior Citizen Concession Card to travel on public transportation (buses, mass rapid transit and light rail transit) at lower fares (TransitLink, n.d.).

Priority Queue Initiative

Since 2015, the Land Transport Authority has implemented Priority Queue to help make older adults' travel journey more comfortable. Implemented at integrated transport bubs and bus interchanges, the initiative provides demarcated waiting/queue zones that enable older adults, pregnant women and people with disabilities to wait comfortably at these areas and board the buses/trains first (Land Transport Authority 2015). The priority queue zones are fitted with seats and design features including tactile to help those who are visually impaired. Additionally, the Land Transport Authority is reviewing and improving wayfinding signs at public transport nodes to make wayfinding more legible and intuitive for older travellers.

Wheelchair-Accessible Buses

Since 2006, all new public buses in Singapore are required to be wheelchair accessible. The entire public bus fleet is expected to be wheelchair-accessible by 2020 (Moogoor, n.d.). To better understand the travel needs of older people, the Public Transport Council and Southwest Community Development Council have introduced a CARE Ride@South West (Caring commuters Assist Readily with Empathy) programme in December 2018 to bring students and older residents together in an attempt to better understand and enrich their public transport travel experience (Public Transport Council 2018).

1.3.3 Housing

The practice is largely directed towards designing and integrating senior facilities and care in and near homes so people can age in place and maintain independent lives. In public housing where over 80% of resident population lives, new building typologies and flat layouts have been designed to meet changing lifestyle needs and provide choice to the ageing population (Yuen and Soh 2017). Purpose-built senior housing and housing modifications are two key strategies to provide age-friendly housing.

Kampung Admiralty

In 2017, residents moved into Singapore's first retirement kampung (Malay word for village), Kampung Admiralty (Sen 2017). The result of a collaborative partnership between government agencies (health, housing, social and community development) and private architecture practice, Kampung Admiralty integrates housing, health and social care within a single vertical development. It comprises two public housing blocks with 104 senior studio apartments. The apartments are designed with age-friendly features to enhance accessibility and safety including handrails, grab bars, ramp, retractable clothes drying rack, induction hobs without open flame, height option for kitchen cabinet to suit wheelchair-bound residents, non-slip flooring, alert alarm system. The integrated housing development also features a medical centre, eldercare and childcare centres, community spaces, supermarket and a hawker centre (Yuen, n.d.). It presents a prototype to develop a supportive network of core amenities, community and eldercare facilities in public housing town to expand at-home and community-based health and social care for ageing in place.

Enhancement for Active Seniors (EASE)

Initially piloted under the City for All Ages project at Marine Parade, flat modification for ageing population is being scaled up nationwide under the Home Improvement— Enhancement for Active Seniors Programme in public housing since July 2012. Under the policy to support changing needs and enable older Singaporeans to pursue their preference to age in place, older residents' homes are retrofitted with senior-friendly features. They can choose to have any combination of the improvements— slip-resistant treatment to flooring in bathrooms/toilets; grab bars; ramps (Housing and Development Board, n.d.). Construction can take up to an average of ten days per flat, depending on the improvements selected and residents need only pay 5–12.5%, depending on flat type, of the programme cost of $2500 per flat (Yuen 2019a). For rental flats, the home improvement items are fully funded by the government.

Monetisation of Housing Asset

Housing design aside, another important practice is to help older residents who are asset-rich but cash-poor to monetise and right size to need-appropriate housing accommodation, usually a smaller dwelling unit. Several support schemes, e.g. sub-letting of rooms, lease buyback, shorter lease, housing bonus are implemented to provide housing choice and help older residents monetise their housing equity and fund retirement (Yuen 2019b).

Continuing Care Precincts

The Housing and Development Board and the Ministry of Health plan to create Continuing Care Housing Precincts where vital facilities are co-located with housing for more convenient use by older adults. This is part of the plan to shift towards care at home and the community by expanding home and community care while providing access to quality and affordable care. The focus is on preventive and responsive care, health not just healthcare. Facilities and services in such precincts include nursing homes, senior care centres, senior activity centres, dementia-friendly communities, day care, home palliative and personal care services (Ministry of Health, n.d.b).

"Smarter" Homes

As part of Singapore's smart nation vision, the Housing and Development Board (HDB) is implementing the HDB Smart Enabled Home Initiative using technology-based solutions to improve daily living in public housing. An example is the elderly monitoring and alert system, which uses motion sensors, elderly profile analysis and emergency detection to detect resident motion and improve home-alone older adults' safety and wellbeing. This and other smart solutions are live-tested with 9000 residents in 3194 flats at Yuhua (2015–2018) to obtain resident feedback including for further refinement.

1.3.4 Social Participation

The focus is on creating opportunity for greater social engagement to help older adults meet people and stay connected with the community.

National Silver Academy

In collaboration with numerous post-secondary education institutions, the National Silver Academy offers subsidised courses to older adults who wish to engage in learning opportunities in their later life (National Silver Academy, n.d.). Older adults can take courses at various universities, polytechnics and institutes of technical education. They range from health and wellness, IT and science, ageing and life skills, humanities, finance and business, and media, arts and design.

Senior Citizens' Executive Committees

Senior Citizens' Executive Committees are older adults' network organised by the People's Association. The committees organise numerous activities for older adults such as cooking and dance classes, exercise and outdoor activities, language courses, among others. The activities are carried out in community clubs. Since 2008, the People's Association (PA) has introduced the PA Wellness Programme to encourage older adults (aged 50 years and above) to care for their own health (see also later discussion under Community Support and Health Services).

1.3.5   Respect and Social Inclusion

The effort highlights the nation's appreciation and respect for the older person who were the pioneers of Singapore's nation building. A British colony since 1819, Singapore attained internal self-rule in 1959 and independence in 1965.

Pioneer Generation Package
This is introduced in 2014 with the establishment of a Pioneer Generation Office[2] to honour and thank Singaporeans who were born on or before 31 December 1949 and obtained citizenship on or before 31 December 1986 for their contribution to Singapore's development. The benefits include subsidised services and medication at polyclinics and specialist outpatient clinics, healthcare insurance premium subsidies, disability assistance, among others (Pioneer Generation Package 2019).

Merdeka Generation Package
This was accorded in June 2019 to Singaporeans born from 1 January 1950 to 31 December 1959, and who became a Singapore citizen on or before 31 December 1996 as an appreciation for their contributions to Singapore. The benefits included active ageing programmes, healthcare, etc. to help the Merdeka generation older adults to stay healthy, active and have assurance over future healthcare costs.

1.3.6   Civic Participation and Employment

The practice is directed towards improving the older person's employment and civic participation opportunities.

Age-Friendly Workplace Practices
In 2017, Tripartite Standards, jointly developed by the Ministry of Manpower, National Trades Union Congress and Singapore National Employers Federation, were introduced to promote good employment practices (Tripartite Alliance Limited, n.d.). There are 8 standards including age-friendly workplace practices that help employers to create a friendly working environment for older adults aged 60 years and above who wish to continue working (Yahya 2018). The Tripartite Alliance for Fair and Progressive Employment Practices offers complimentary Tripartite Standards Clinic to help employers become adoption ready.

RSVP Singapore
Started in 1998, RSVP Singapore or the Organisation of Senior Volunteers is an institution of public character and the national centre of excellence for senior volunteerism (RSVP Singapore, n.d.). Its focus is on building relationships, innovation,

---

[2]The Pioneer Generation Office was merged with the Agency for Integrated Care and renamed the Silver Generation Office in April 2018. It is presently managed by the Ministry of Health in an effort to streamline care for older adults.

integrity and service to others as well as sharing experiences. The vision is to make every senior a volunteer through activities such as the Enriching Lives of Seniors Programme where volunteers befriend the socially isolated older adults; Mentally Disadvantaged Outreach Programme in which volunteers befriend mental health patients; the Mentoring Programme in which volunteers mentor schoolchildren; MyBuddy Programme in which volunteers befriend socially isolated older adults and those that were recently discharged from hospitals; Episodic Volunteering Programme where potential volunteers can try out short-time volunteering.

1.3.7 Communication and Information

The emphasis is on the dissemination of information and resources to help the older person get around and know where to get help if needed.

Singapore Silver Pages
Introduced in 2011, this initiative was developed by the Agency for Integrated Care as a one-stop resource portal for information on community care for older adults, caregivers and care decision makers to help them make informed care decisions (Singapore Silver Pages, n.d.). The information canvasses social care, healthcare, mental health, care services and financial help.

1.3.8 Community Support and Health Services

The practice is largely directed towards building senior-friendly communities through the development of community support and health services to encourage healthy lifestyles.

"You Can Get Moving"
The exercise campaign, You Can Get Moving was launched in 2017 by the Health Promotion Board to encourage people aged 50 and above to improve their strength, balance and flexibility (Health Promotion Board 2017). The exercise programme, 7 Sit-Down Exercises, is designed to be easy to do with minimal equipment (a chair and a towel). Older adults can perform these exercises at home or outside. Instructions and safety tips are presented through posters and videos in the nation's 4 major languages: English, Mandarin, Malay and Tamil, for easy access and appreciation by all older adults.

Project Silver Screen
Launched in 2018, the project promotes good health through regular health screening to monitor the functional ability of all Singaporeans aged 60 years and older in 3 key areas: vision, hearing and oral health. The screenings are organised by the Health

Promotion Board and conducted in easily accessible places such as Community Centres, Residents' Committees or other public spaces that are frequented by older adults. The screening cost ranges from S$0–5 (Health Hub, n.d.).

Active Ageing Programmes
In order to keep older adults active, healthy and socially engaged, various activities are organised for them in their neighbourhoods under the Ministry of Health Active Ageing Programme. The group activities include exercise classes like Zumba Gold, K-Pop fitness, stretch band exercises, low impact aerobics, health workshops, healthy cooking classes and other social activities such as karaoke and Rummy-O sessions (Ministry of Health, n.d.a).

Wellness Programme
The People's Association organises wellness programmes for older adults aged 50 years and older to maintain and improve their health (People's Association 2019). The emphasis is on staying mentally, physically and socially active through health screenings, physical activities and social networking in Community Clubs, Residents' Committees and Neighbourhood Committees. In 2018, the PA Wellness Programme has been extended from community clubs to 200 residents' committees. The target is to reach 400 residents' committees by 2020. The Wellness Time@RC provides a programme of regular active ageing activities for older residents in their communities.

Community Network for Seniors
Piloted in 2016, this network comprises various stakeholders such as voluntary welfare organisations, grassroots organisations, regional health systems and government agencies. The goals include conducting preventive health home visits, encouraging older adults to stay physically and socially active, providing basic healthcare services to neighbourhoods via preventive health screenings, providing community support through befriending schemes, and providing senior-centric help (Centre for Liveable Cities and the Seoul Institute 2019). The current plan is to expand the Community Network nationwide by 2020.

Senior Activity Centres
Since 1995, Senior Activity Centres have been established as neighbourhood drop-in centres, usually located on the ground floor of public housing blocks. The target group is the lower income older adult. The centres offer recreational and social activities as well as support services such as emergency response, information or referral for care support (Health Promotion Board 2017).

Active Ageing Hubs
Located in public housing neighbourhoods, these are large integrated hubs that provide both active ageing and care services for active and ambulant older adults as well as day care, day rehabilitation and assisted living services e.g. housekeeping, grocery shopping for frail older adults. Designed to be one-stop centres, Active Ageing

Hubs also organise activities such as healthy cooking classes, fall-risk and wellness screenings, advanced care and retirement planning programmes, and intergenerational programmes that bring children and older adults together (Health Hub, n.d.). Currently, 6 have been developed out of a total of 10 active ageing hubs planned by 2020.

1.4 Emerging Research

Compared to international literature, Singapore's research on age-friendly city is largely nascent. The majority of the local studies are focused on ageing in place and housing as well as social participation among older adults. As older adults spend most of their time and daily activities near their home, it is of utmost importance to understand the impact of not just the dwelling unit but also the neighbourhood built environment on older people.

1.4.1 Housing, Built Environment, Transportation

Studies on age-friendly housing in Singapore generally work in the notion of ageing in place, which highlights not only the importance of right-sized dwelling unit and living arrangements but also a familiar neighbourhood for older people (Yuen and Soh 2017). Since the majority of the residential development is public housing, most of the studies are understandably about this context. Few have studied private housing residents, suggesting an opportunity for further research. Yuen (2019c) and her team in their survey of both public and private housing residents have examined their willingness to age in place and the determinants of age-friendly housing, which included good location, house ownership, accessible services, senior-friendly features within homes and ease of home maintenance. Most of the studies are cross-sectional and there is scope for more longitudinal and panel studies to track attitude and behavioural change over time.

   Another potential area for further research is on age-friendly neighbourhood environment. There is growing literature on older Singaporeans' mobility (e.g. Koh et al. 2015), psycho-social wellbeing (Chong et al. 2015), subjective and objective measures of neighbourhood environments and relationships with transportation physical activity (Nyunt et al. 2015) that support the integration of age-friendly features into the planning and design of neighbourhood environments, particularly the availability and accessibility of amenities. Again, many of these studies are confined to public housing neighbourhoods. They suggest that older Singaporeans have social engagement and leisure activities at commercial amenities, service centres, open spaces and other public spaces (Cho et al. 2018; Chong et al. 2015; Yuen 2019c). With few exceptions, these studies generally do not examine the connection between the built environment and health. The principles of age-friendly design interventions are seldom investigated even though barrier-free living environment, residential density, diversity of land use mix and aesthetic environment are found to be facilitators of walking behaviour among older Singaporeans (Nyunt et al. 2015; Koh et al. 2015).

Among the many research centres on population ageing (there is at least one in every major university), the Lee Li Ming Programme in Ageing Urbanism at the Lee Kuan Yew Centre for Innovative Cities, Singapore University of Technology and Design undertakes multi-disciplinary ageing research with a focus on the built environment. Since 2014, the programme has undertaken evidence-based research to develop a holistic understanding of the connection between the built environment, health and quality of life that can inform the planning and design of age-friendly neighbourhoods and communities. According to its webpage (https://lkycic.sutd.edu. sg/research/ageing-urbanism/), it has recently completed a multi-disciplinary, multi-methods translational research on Innovative Planning and Design of Age-friendly Neighbourhoods in Singapore.

Another recent centre for built environment research is the Centre for Ageing Research in the Environment (CARE) at the School of Design and Environment, National University of Singapore. It emphasises research and education related to ageing, healthcare and the environment with a focus on high-rise, high-density urban environments. Yet another is the Ageing Research Institute for Society and Education (ARISE) at the Nanyang Technological University. Launched in 2016, ARISE is a pan-university institute that supports multidisciplinary and interdisciplinary ageing-related research with focus on ageing medicine, social integration and education, ageing in place, and care and lifestyle enhancement. With the fast emergence of research centres on the built environment, we can look forward to more research in the coming years.

1.4.2   Social Participation, Respect and Social Inclusion

International research has suggested that social participation helps to strengthen wellbeing, prevent social isolation and reduce death risk among older adults (Carver et al. 2018; Wu and Li 2018). For Singapore, many of the studies are conducted on older residents living in high-rise public housing. For example, Aw et al. (2017), using mixed methods, have identified and explained a continuum of social participation among 109 older adults (aged 55 years and older) in a public housing neighbourhood, which is part of the project site of the City for All Ages Community for Successful Ageing at Whampoa. The proposed continuum or ladder of older person's social participation ranges from marginalisation and exclusion to comfort-zoning alone, seeking consistent social interactions, expanding one's social network and social engagement (volunteering and helping others), and the factors that could help the older person move up the continuum of social participation include financial security, good health status and management, psychological adjustment to social interactions, family support and integration.

In a separate study, Cho et al. (2018) report that older adults are significantly engaged in socialisation and leisure activities at commercial amenities where they meet friends and neighbours, watch people, and enjoy the environment and quiet time. Chong et al. (2015) have suggested that older Singaporeans engage in social participation and form friendships at both formal social service centres and informal public spaces. They further suggested that active participation in activities leads to more satisfaction with the housing environment, which could contribute to greater sense of social cohesion and self-efficacy, and lower risk of depression.

In 2019, the Centre for Ageing Research and Education (CARE), Duke-NUS Medical School released the key findings from its Transitions in Health, Employment, Social Engagement and Inter-generational Transfers in Singapore Study (SIGNS Study)-I (Chan et al. 2018). Conducted in 2016–2017 (first wave), the nationally representative longitudinal study interviewed 4549 community-dwelling older Singaporeans aged 60 years and older about their demographics, income and financial adequacy, physical and functional health (e.g. chronic diseases, self-rated health, body mass index, hand grip strength), psychological health, health behaviours (e.g. health screening, physical activity, smoking, prescription medication use and adherence, healthcare utilisation), social engagement (e.g. living alone, social network outside the household, attendance in social activities, loneliness), provision and receipt of transfers, work and retirement, lifelong learning, volunteering and monetary donation. The findings revealed that 9% of older Singaporeans lived alone and those with weaker social networks reported feeling lonely and have more depressive symptoms. The barriers to social connectedness included individual instrumental activities of daily living limitations and environmental access. Aside from software development, e.g. activity programming, the report highlighted the need to improve/redesign the neighbourhood environment to enable greater ease of mobility among the older population.

1.4.3 Community Support and Health Services

To better understand the development towards healthy living habits, a small but growing number of studies have been conducted on the community health-promoting practices in Singapore. For example, Kailasam et al. (2019) have conducted an assessment of the prevalence of community health-promoting practices using a composite health promotion scoring system to examine 5 domains: community support and resources; healthy behaviours; chronic conditions; mental health; and common medical emergencies. The conclusion is that community health-promoting practices are not adequate in Singapore, especially in the areas related to community support and resources, chronic disease prevention and management, and efforts to support and nudge the adoption of healthy behaviours.

Low et al. (2017) proposed a methodology to evaluate the effectiveness of the integrated community of care programme, a novel care model that integrates hospital-based care with health and social care in the community for high-risk individuals living in socially deprived communities. The team recruited patients aged 60 years and older, staying in public rental housing and would use a mix of quantitative and qualitative methods to study user experience over a 2-year period, July 2016 to June 2018. They anticipate that their findings could inform policymakers on the feasibility of shifting care from a hospital centric system to an integrated community centric system for high-risk communities, and implementation of future iterations of this care model.

As home and community care become more important in Singapore's healthcare system, there is opportunity for more research on community support and health services. Since 2015, the Geriatric Education and Research Institute, established as a national entity under the Ministry of Health, has begun to develop, coordinate and implement initiatives to strengthen geriatric education, research and service planning in the clinical and health aspects of ageing in Singapore. Its focus areas include health services and policy research; seniors care delivery and health outcome; frailty identification, prevention and management; clinical geriatrics and end of life care.

1.5   Concluding Remarks

Unlike many cities, Singapore started to focus on population ageing issues when the population was still predominantly young. Understandably, at the start in the 1980s, the policy focus was largely economic-centric. The desideratum was on developing the workforce and projecting the challenge of an ageing labour force as Singapore continued to transform its economic growth with industrial revolution 2.0. Moving into the 1990s with a greater focus on quality of life consideration, a holistic, multi-agency approach was adopted, expanding policy purview on ageing in place, community service, information provision and transportation issues. The built environment emphasis continued into the 2000s with renewed impetus, with deepening focus on integrating healthcare, improving the neighbourhood built environment and providing support for caregivers. From 2010 onwards, this emphasis was turned to create ample opportunities for physical and social interaction through a slew of policies targeted at programming activities and creating space that is conducive for such purposes. An increasing emphasis in policy development is partnership and engagement with stakeholders to better understand and address the needs of the ageing population.

Over the decades, Singapore has implemented numerous programmes and activities pertaining to all 8 domains of the WHO Age-friendly Cities framework even though it is not a member of the WHO global network. Guided by a vision of Successful Ageing for Singapore and an ambitious national plan, the latest (Ministerial Committee on Ageing 2016) effort included more than 70 initiatives in 12 areas, from health and wellness to learning, volunteerism, employment, housing, transport, public spaces, respect and social inclusion, retirement adequacy, healthcare and aged care, protection for vulnerable, and research. Increasingly, the effort is to involve a broad range of community stakeholders including voluntary organisations, faith-based organisations and the private sector in vision implementation. Singapore has put in place a large and growing body of activities and programmes that aim to improve the built environment, social connections and health of older adults. The overarching goal is to promote active and healthy ageing in place.

In contrast, research remains an area for further development. Since the inclusion of research in the national plan for successful ageing, more research funding has been made available. An example is the Silver Age Research Programme funded by the Land and Liveability National Innovation Challenge under the Prime Minister's Office. Managed by the Urban Redevelopment Authority in collaboration with the Ministry of Health and the Housing and Development Board.

The ambit of the Silver Age Research Programme covers targeted research that would improve the way Singapore plans for an inclusive and age-friendly city, create physical and social environments that promote active ageing, respond to evolving needs and aspirations of older people, and provide care services and infrastructure support. More research opportunity awaits.

References

Ageing Planning Office (n.d.). Ageing in place in Singapore. Retrieved from: http://www.gs.org. sg/sg50conference/pdf/s4-1.pdf. Accessed on 30 July 2019.

Aw, S., Koh, G., Oh, Y. J., Wong, M. L., Vrijhoef, H. J. M., Harding, S. C., et al. (2017). Explaining the continuum of social participation among older adults in Singapore: From 'closed doors' to active ageing in multi-ethnic community settings. *Journal of Aging Studies, 42,* 46–55.

Carver, L. F., Beamish, R., Phillips, S., & Villeneuve, M. (2018). A scoping review: Social participation as a cornerstone of successful aging in place among rural older adults. *Geriatrics, 3,* 75. https://doi.org/10.3390/geriatrics3040075.

Centre for Liveable Cities, & The Seoul Institute (2019). Age-friendly cities: Lessons from Seoul and Singapore. Retrieved from: https://www.clc.gov.sg/docs/default-source/books/book-age-friendly-cities.pdf. Accessed on 26 July 2019.

Chan, A., Malhotra, R., Manap, N., Ting, Y. Y., Visaria, A., Cheng, G. H., et al. (2018). *Transitions in health, employment, social engagement and intergenerational transfers in singapore study (The SIGNS Study)-I: Descriptive statistics and analysis of key aspects of successful ageing.* Singapore: Centre for Ageing Research and Education, Duke-NUS Medical School.

Cho, M., Ha, T. M., Lim, Z. M. T., & Chong, K. H. (2018). "Small places" of ageing in a high-rise housing neighbourhood. *Journal of Aging Studies, 47,* 57–65.

Chong, K. H., Yow, W. Q., Loo, D., & Patrycia, F. (2015). Psychosocial well-being of the elderly and their perception of matured estate in Singapore. *Journal of housing for the elderly, 29*(3), 259–297.

Department of Statistics. (2019). Population Trends, 2018. Retrieved from https://www.singstat. gov.sg/-/media/files/publications/population/population2018.pdf. Accessed on 30 July 2019.

Doran, C. F., & Dixon, C. (1991). South East Asia in the world-economy. Cambridge: Cambridge University Press.

Graham, E. (1995). Singapore in the 1990s: Can population policies reverse the demographic transition? *Applied Geography, 15*(3), 219–232.

Health Hub. (n.d.). See, hear and eat better. Retrieved from https://www.healthhub.sg/programmes/ 144/functional-screening. Accessed on 12 July 2019.

Health Promotion Board. (2017). New exercise campaign to encourage active living among seniors, including those with reduced mobility. Retrieved from https://www.hpb.gov.sg/article/new-exercise-campaign-to-encourage-active-living-among-seniors-including-those-with-reduced-mobility. Accessed on 12 July 2019.

Housing and Development Board.. (n.d.). Enhancement for active seniors (EASE). Retrieved from https://www.hdb.gov.sg/cs/infoweb/residential/living-in-an-hdb-flat/for-our-seniors/ease. Accessed on 23 July 2019.

Kailasam, M., Hsann, Y. M., Vankayalapati, P., & Yang, K. S. (2019). Prevalence of community health-promoting practices in Singapore. *Health Promotion International, 34,* 447–453.

Koh, P., Leow, B., & Wong, Y. (2015). Mobility of the elderly in densely populated neighbourhoods in Singapore. *Sustainable Cities and Society, 14,* 126–132.

Land Transport Authority. (n.d.). Silver zones. Retrieved from https://www.lta.gov.sg/content/ltaweb/en/roads-and-motoring/projects/road-and-commuter-facilities/silver-zones.html. Accessed on 23 July 2019.

Land Transport Authority. (2013). Longer green man time for elderly and disabled pedestrians at 239 more pedestrian crossings. Retrieved from https://www.lta.gov.sg/apps/news/page.aspx?c=2&id=830c5cd6-f3cf-4624-b35b-f0f84dbb441d. Accessed on 23 July 2019.

Land Transport Authority. (2015). Factsheet: creating a senior-friendly transport system. Retrieved from https://www.lta.gov.sg/apps/news/page.aspx?c=2&id=d34b11e6-6b67-4333-b6d2-ad76e3487525. Accessed on 23 July 2019.

Low, L. L., Maulod, A., & Lee, K. H. (2017). Evaluating a novel integrated community of care (ICoC) for patients from an urbanised low-income community in Singapore using the participatory action research (PAR) methodology: A study protocol. *British Medical Journal Open, 7,* e017839.

Ministerial Committee on Ageing. (2016). *Action Plan for Successful Ageing.* Singapore: Ministry of Health.

Ministry of Health (n.d.a). Active ageing programmes. Retrieved from https://www.moh.gov.sg/ifeelyoungsg/how-can-i-age-actively/stay-healthy/active-ageing-programmes. Accessed on 11 July 2019.

Ministry of Health (n.d.b). Live in friendlier homes. Retrieved from https://www.moh.gov.sg/ifeelyoungsg/how-can-i-age-in-place/live-in-friendlier-homes. Accessed on 22 July 2019.

Ministry of Health. (1984). *Report of the Committee on the Problems of the Aged.* Singapore: Ministry of Health.

Ministry of Social and Family Development. (2006). *Report of The Committee on Ageing Issues, 2006.* Singapore: Ministry of Social and Family Development.

Moogoor, A. (n.d.). Public transport systems for ageing populations. Retrieved from: https://lkycic.sutd.edu.sg/research/resources/ Accessed on 24 September 2019.

National Parks Board. (2019). Therapeutic gardens. Retrieved from https://www.nparks.gov.sg/gardens-parks-and-nature/therapeutic-gardens. Accessed on 23 July 2019.

National Silver Academy. (n. d.). About National Silver Academy. Retrieved from https://www.nsa.org.sg/about.php. Accessed on 12 July 2019.

Nyunt, M. S. Z., Shuvo, F. K., Eng, J. Y., Yap, K. B., Scherer, S., Hee, L. M., et al. (2015). Objective and subjective measures of neighborhood environment (NE): Relationships with transportation physical activity among older persons. *International Journal of Behavioral Nutrition and Physical Activity, 12*(1), 108.

People's Association. (2019). Active ageing. Retrieved from: https://www.pa.gov.sg/our-programmes/active-ageing. Accessed on 25 July 2019.

Pioneer Generation Package. (2019). What are the benefits? Retrieved from: https://www.pioneers.sg/en-sg/Pages/Home.aspx. Accessed on 22 July 2019.

Public Transport Council. (2018). A bus ride across generations to foster a caring commuting experience. Retrieved from: https://www.ptc.gov.sg/newsroom/news-releases/newsroom-view/a-bus-ride-across-generations-to-foster-a-caring-commuting-experience. Accessed on 24 September 2019.

RSVP Singapore. (n. d.). Vision and mission. Retrieved from: https://rsvp.org.sg/vision-mission/. Accessed on 25 July 2019.

Scharlach, A. (2012). Creating aging-friendly communities in the United States. *Ageing International, 37*(1), 25–38.

Sen, N. J. (2017, August 12). Residents collect keys at Singapore's first "retirement kampung". *The Straits Times*. Retrieved from: https://www.straitstimes.com/singapore/residents-collect-keys-at-singapores-first-retirement-kampung. Accessed on 22 July 2019.

Singapore Silver Pages. (n. d.). Welcome to the Singapore Silver Pages. Retrieved from: https://www.silverpages.sg/AboutUs. Accessed on 12 July 2019.

Song, S. (n.d.). Safer streets for seniors in Singapore. Retrieved from: https://lkycic.sutd.edu.sg/research/resources/ Accessed on 22 September 2019.

TransitLink. (n.d.). Senior citizen concession card. Retrieved from: https://www.transitlink.com.sg/PSdetail.aspx?ty=art&Id=144#4. Accessed on 23 July 2019.

Tripartite Alliance Limited. (n.d.). Age-friendly workplace practices. Retrieved from: https://www.tal.sg/tafep/getting-started/progressive/tripartite-standards#age-friendly. Accessed on 12 July 2019.

World Health Organization. (2007). *Global age-friendly cities: A guide*. Geneva: World Health Organization.

Wu, J., & Li, J. (2018). The impact of social participation on older people's death risk: An analysis from CLHLS. *China Population and Development Studies, 2*(2), 173–185.

Yahya, Y. (2018, April 27). New standard to help companies adopt age-friendly workplace practices. *The Straits Times*. Retrieved from: https://www.straitstimes.com/singapore/new-standard-to-help-companies-adopt-age-friendly-workplace-practices. Accessed on 12 July 2019.

Yuen, B. (n.d.). Senior public housing in Singapore: Kampung Admiralty. Retrieved from: https://lkycic.sutd.edu.sg/research/resources/. Accessed on 24 September 2019.

Yuen, B. (2019a). Adapting public housing to age in place in Singapore In A. P. Lane (Ed.), *Urban environments for healthy ageing*. Abingdon: Routledge

Yuen, B. (2019b). Moving towards age-inclusive public housing in Singapore. *Urban Research and Practice, 12*(1), 84–98.

Yuen, B. (2019c). *Ageing and the built environment in Singapore*. Berlin: Springer.

Yuen, B., & Soh, E. (2017). *Housing for older people in Singapore: An annotated bibliography*. Berlin: Springer.

Chapter 2
Annotated Bibliography

2.1 Overview

This annotated bibliography contains not just Singapore literature (both published and grey literature) but also published literature on other cities, covering Asia-Pacific, North America and Europe. The intent is to give a quick pulse on the kind of age-friendly city research that is taking place in these places. The key search terms used for the annotation compilation included age/senior/elder-friendly city/neighbourhood/precinct, World Health Organisation (WHO) Age-friendly Cities framework/model/programme/initiative, WHO 8 age-friendly city/age-friendliness domains. The literature review is not exhaustive. Only publications in English language have been included. However, we recognise the presence of a growing body of literature on age-friendly cities in other languages that have been inaccessible to us due to language barrier. Perhaps, future research could examine this literature. As research continues, the publication landscape will continue to evolve.

Asia-Pacific

The Asia-Pacific literature on age-friendly cities and communities spans 10 countries: Australia, China (and Hong Kong SAR), India, Indonesia, Japan, Malaysia, New Zealand, South Korea, Taiwan and Thailand. Several common underlying themes can be identified in the age-friendly discourse between groups of countries: China, Hong Kong SAR and Taiwan; Japan and South Korea; Australia and New Zealand; Malaysia and Thailand. These themes can be attributed to similarities among features, including but not limited to, the mode of governance, societal contexts and developmental state.

Australia and New Zealand

The age-friendly city discourse in Australia and New Zealand highlights the need to consider the heterogeneity of older adults' preferences to enable ageing in place. An interdisciplinary approach to policymaking emerged as a consistent recommendation across multiple studies in both countries. A somewhat similar but distinct approach to the advancement of age-friendly discourse may be observed in both countries. While age-friendly literature in Australia focuses on the conceptual advancement of age-friendly discourse such as exploring ideas of intergenerational space and evaluation of assessment indicators, New Zealand literature examines the application of ecological models to the age-friendly city framework.

China, Hong Kong SAR and Taiwan

There is an active focus on the themes of urban governance, built environment and regional disparity. Specifically, accessibility to parks and opportunities for physical activity are highlighted repeatedly in the literature. The emphasis on green space might be due to an urgent need to alleviate the effects of high-density urban living. The direction of age-friendly discourse differs with governance structure. In China, age-friendly city policies are primarily top-down with an emphasis on addressing widespread regional inequality while Taiwan and Hong Kong SAR adopt a consensus-building approach to the development and implementation of age-friendly policies.

Japan and South Korea

Contextualised community resilience to urban pressures is a common theme that has emerged consistently across the age-friendly city discourse in Japan and South Korea. Studies have explored the role of social capital as a driving force in creating economic and structural models that help to sustain communities facing gentrification pressures. Both countries adopt and build upon the WHO Age-friendly Cities framework in different ways. A huge emphasis is on the development of contextually relevant age-friendly indicators in South Korea while the Japanese literature elaborates on tools to study the WHO Age-friendly Cities domains to help augment policymaking.

Malaysia and Thailand

The Malaysia and Thailand literature shares a common theme—the development of the built environment. Empirical literature in both countries stresses the importance of the creation of barrier-free environments to enable active mobility and access to essential amenities for older adults.

North America

The age-friendly city research review for North America covers the United States and Canada where some of the cities are members of the WHO Global Network of Age-Friendly Cities and Communities. Research topics canvass all 8 domains of the WHO Age-friendly Cities framework as well as some other age-friendly domains based on the re-conceptualisation of age-friendly city according to local context. Most studies concentrate on exploring age-friendliness from the perspective of older adults.

United States

Research studies in the United States cover the development of the WHO Age-friendly Cities project and related assessment as well as age-friendly initiatives development of elaboration, evaluation and linkage between policy, knowledge and practice. Studies are conducted in both urban and rural geographic study sites, and on naturally occurring retirement communities (NORCs), an age-friendly neighbourhood that is uniquely American. A diverse methodology has been adopted including participatory research such as World Café, conferences, workshops, etc. that involve older adults and other stakeholders.

Canada

Research on age-friendly neighbourhoods in Canada has more emphasis on age-friendliness evaluation and initiative assessment to identify the enablers and challenges for ageing in place. Studies are conducted in different geographical areas with various community characteristics including vulnerable rural communities. Some research has drawn on or re-conceptualised the WHO Age-friendly Cities framework while others focused on its individual age-friendly domains. In addition, much of the age-friendly research in Canada has adopted community-based participatory methodology to obtain insights on the perception and understanding from older adults, highlighting the participatory role of older adults. Findings from age-friendly studies in Canada generally include implication for policy or planning in relation to age-friendly neighbourhoods.

Europe

The European literature on age-friendly city spans 12 countries: Belgium, Finland, France, Germany, Ireland, Italy, the Netherlands, Portugal, Slovenia, Spain, Sweden and the United Kingdom. A common theme across the majority of these countries is the investigation of older adults' perceptions of the built environment/their neighbourhood and their effects on older adults' lived experiences.

United Kingdom

Studies in the United Kingdom mostly focus on the relationship between the built environment or the perceptions of the built environment and older adults' health such as wellbeing, walking, and quality of life. Besides this, the research topics explore the meaning or conceptualisation of age-friendliness and the formation of age-friendly policies. Other research issues include inclusivity and social responsibility in connection with age-friendly initiatives.

Other European countries

Research in other European countries mostly focuses on older adults' perceptions of their neighbourhoods and its association with different aspects of older adults' lives such as walking for transportation or different types of health. The studies also explore reasons behind people's relocation in later life. Several studies address both the European Healthy Cities Network as well as the implementation of the WHO Age-friendly Cities model in European cities. Topics of ageism and conceptualisation of ageing in place are also investigated.

2.2 Singapore[1]

2.2.1 Books and Journal Articles

Addae-Dapaah, K., & Wong, G. (2001). Housing and the elderly in Singapore— Financial and quality of life implications of ageing in place. *Journal of Housing and the Built Environment, 16*(2), 153–178.

This paper examined the home modification options and related financial support in public housing for ageing population in Singapore. Archival literature review on senior housing, financial gerontology and demography was complemented with empirical research. Two surveys were conducted with 1500 and 751 older Singaporeans aged 50 years and older to collect information on their housing needs and residential mobility, and preference and affordability for senior-friendly features respectively. Results suggested that the practical strategy for addressing housing problems and improving quality of life of older adults in Singapore might be home modification instead of specialised elderly housing. Self-help and government intervention were identified as two key strategies to improve senior-friendly home modification.

Centre for Liveable Cities. (2019). *Age-friendly cities: Lessons from Seoul and Singapore.* **Singapore: Centre for Liveable Cities**.

This book analysed, compared and presented facts, achievements, case studies and research on ageing landscape and age-friendly cities related themes of two Asian cities: Seoul and Singapore. The Singapore discussion detailed the challenges, current efforts and research on age-friendly neighbourhoods in Singapore while the Seoul case studies examined efforts to address ageing in place including for people living with dementia.

Chan, A. (2008). Social policies for the aged in Singapore. In K. F. Lian & C. K. Tong (Eds.), *Social policy in post-industrial Singapore.* **Boston: Brill.**

Amidst rapid population ageing, this paper highlighted the key ageing policy aspects in Singapore and evaluated their effectiveness. Arguing that the current model of family dependence on public financial and social support was unsustainable, recommendations were raised to improve financial independence, provide infrastructure with increasingly modernist living aspirations and augment formal systems of support (e.g. Central Provident Fund savings for retirement, nursing homes). The author

---

[1]The grey literature—unpublished dissertation and news reports on Singapore were compiled with the help of SUTD Library.

concluded that as the number of older Singaporeans continued to increase, early governmental policies that could facilitate the independence for older adults, socially and financially, were critical to ameliorating future effects of population ageing in Singapore.

Chan, A., & Matchar, D. B. (2015). Demographic and structural determinants of successful aging in Singapore. In S. T. Cheng, I. Chi, H. H. Fung, L. W. Li, & J. Woo (Eds.), *Successful aging: Asian perspectives* (pp. 65–79). Dordrecht: Springer Netherlands.

This chapter presented Singapore as a case study on the determinants for successful ageing. The determinants covered demographic change and its impact on health and service needs, economic, social and cultural issues; state policy for social support and health care of older population, especially policies for older women and the implication of cohort differences. For instance, the baby boomer generation born between 1947 and 1964 were expected to have different characteristics from their older counterparts such as more retirement savings, higher education and income, which might lead to a different conceptualisation of successful ageing in Singapore. The authors acknowledged the importance of innovation in supporting future generations of older adults and their needs.

Cho, M., Ha, T. M., Lim, Z. M. T., & Chong, K. H. (2018). "Small places" of ageing in a high-rise housing neighbourhood. *Journal of Aging Studies, 47*, 57–65.

This study examined the spaces that facilitated mundane and everyday social interaction among older residents in high-rise public housing neighbourhood in Singapore. Using survey, mapping and resident engagement, data on resident demographics, activities and perception of neighbourhood spaces were collected from 610 residents (15.4% older adults were aged 65 years and above) in Yuhua estate. Logistic regression analysis revealed that older adults were significantly engaged in socialisation and leisure activities at commercial amenities, e.g. coffee shops and neighbourhood shops. The findings showed that a compact living environment would entail better accessibility as well as tighter regulations over space usage. The authors acknowledged that a top-down planning approach that was detached from residents' life-stages and daily life was more likely to end up prioritising efficiency to liveability.

Chong, K. H., Yow, W. Q., Loo, D., & Patrycia, F. (2015). Psychosocial well-being of the elderly and their perception of matured estate in Singapore. *Journal of Housing for the Elderly, 29*(3), 259–297.

This study explored the relationships between age-friendly community and older adult's psychosocial wellbeing in Singapore. Based on the concept of 'ageing friendliness', the perceptions and needs of 5Cs (continuity, compensation, connection, challenge, and contribution), demographic, subjective physical health, Montreal Cognitive Assessment, and Geriatric Depression Scale-15 (GDS-15) were collected through quantitative surveys and qualitative interviews with 33 older adults aged

55 years and older living in Bukit Merah public housing estate. Results suggested that older Singaporeans socially participated and formed friendships at both formal social service centres and informal public spaces. Active participation in activities led to more satisfaction with the housing environment, which could contribute to a greater sense of social cohesion and self-efficacy, and lower risk of depression. Older Singaporeans were generally satisfied with the physical infrastructure in their neighbourhood. The authors highlighted the importance of comprehensive, integrated urban design to facilitate physical activities, social interactions and active ageing in order to enhance older adults' psychosocial wellbeing.

Chong, K. H., & Kang, F. I. (2018). *Second beginnings: Senior living redefined.* **Singapore: Lien Foundation.**

The book was commissioned by the Lien Foundation to study the current state and new concepts of senior living in Singapore. Ten new senior living typologies were proposed to address the various needs of the growing ageing population. These included designs of community spaces where older people would work, play and receive care, a mobile senior activity centre on wheels, and various models of assisted living in which senior housing was fitted with age-friendly features and eldercare support. Older people were actively engaged in the design process to provide inputs about their everyday lives and interactions with spaces.

Koh, P., Leow, B., & Wong, Y. (2015). Mobility of the elderly in densely populated neighbourhoods in Singapore. *Sustainable Cities and Society, 14*, 126–132.

This study identified infrastructure features, socio-demographic and situational variables that promoted mobility and created senior-friendly environment for older adults in Singapore. Through a perception interview survey, 168 older adults aged 55 years and above reported their estimated walking duration, quality of perceived surrounding environments (e.g. security, detour, etc.), and socio-demographic and situational variables. Geographical information system was used to capture their movement information. The analysis showed that older adult fallers sustained lower mobility level and walking duration. The results supported barrier-free provisions and senior-friendly fittings and features for ageing in place. The authors recommended further research on social travel demand of older adults.

Lane, A. P., Wong, C. H., Močnik, Š., Song, S., & Yuen, B. (2019). Association of neighborhood social capital with quality of life among older people in Singapore. *Journal of Aging and Health*. https://doi.org/10.1177/0898264319857990.

This paper assessed the association between neighbourhood social capital—cognitive (social cohesion) and structural (individual participation in groups and organisations)—and quality of life among Singaporean older adults. The study sample consisted of 981 community-dwelling adults aged 55 years and above in three public housing neighbourhoods who completed a survey questionnaire administered as part of the research project, 'Innovative Planning and Design of Age-Friendly Neighbourhoods in Singapore'. Adopting multilevel models on this cross-sectional survey data, quality of life among older residents was found to be higher in neighbourhoods where residents actively participated in groups and organisations and are socially cohesive.

Nyunt, M. S. Z., Shuvo, F. K., Eng, J. Y., Yap, K. B., Scherer, S., Hee, L. M., Chan, S. P. & Ng, T. P. (2015). Objective and subjective measures of neighborhood environment (NE): Relationships with transportation physical activity among older persons. *International Journal of Behavioral Nutrition and Physical Activity,* *12*(1), 108.

This study assessed the associations between objective and subjective neighbourhood environment and transportation physical activity among community-dwelling older adults in Singapore. Trained research nurses conducted face-to-face interviews with 402 older adults aged 55 years and above in 3 public housing precincts with residential length of 5 years and more from the Singapore Longitudinal Ageing Study Wave 2. The interview collected information on their socio-economic and physical health status, walking frequency for transportation purpose, subjective and objective neighbourhood environment measures (using an adapted version of the Neighbourhood Environment Walkability Scale (NEWS) and Geographical Information System respectively). Walkability and accessibility indexes were computed, and multiple regression models applied to study the association of subjective and objective measures of the built environment characteristics with transportation physical activity. The findings showed that subjective residential density, diversity of land use mix, aesthetic environment, and objective proximity to amenities and facilities were significantly correlated with higher frequency of walking for transportation purposes.

Rogerson, A., & Stacey, S. (2018). Successful ageing in Singapore. *Geriatrics,* *3*(4), 81.

This paper reviewed Singapore's national and regional proactive measures and policy changes to respond to rapid population ageing. The measures were in the domains of Employment, Lifelong Learning, Volunteerism, Communities of Care, Intergenerational Harmony and Respect for Seniors, Geriatric Medicine, Care and Research, and Transport and Urban Planning. In addition, the authors presented the example of Sengkang General Hospital in north-eastern Singapore, which had implemented a community-based programme on frailty screening and post-screening care.

Yap, M. T. (2008). Singapore's response to an ageing population. In H. G. Lee (Ed.), *Ageing in Southeast Asia and East Asia: Family, Social Protection and Policy Challenges* **(pp. 66–87). Singapore: Institute of South-East Asian Studies.**

This book chapter discussed the development of policies and programmes relating to ageing population in Singapore over time. The Singapore government had worked towards affordable, sustainable solutions, which encouraged social cohesion to ensure successful ageing for Singaporeans. A major focus was on financial security, healthcare and social integration policies and programmes, recommending regular review in these areas to ensure effectiveness of relevant policies as well as future studies to understand the evolving needs and requirements of older people.

Yuen, B. (Ed.) (2019). *Ageing and the built environment in Singapore.* **Cham: Springer**.

This book shared the research methodologies and findings from the research project, Understanding the Changing Needs of Singapore's Older Population that studied two key activity spaces—live and play among older adults aged 55 years and older in Singapore. The results from qualitative and quantitative research including survey (public and private housing residents), focus group discussion, key informant interview, on-site observation and community design workshop illustrated older adults' preferences and perceptions about their living and recreation arrangements, residential mobility and satisfaction, ageing and growing old in Singapore. The project also developed tools like housing audit (using the WHO age-friendly housing features and universal design principles), walk and talk survey, see and snap photo interview and real-time mobility mobile app on android phone for data collection.

Yuen, B., & Soh, E. (2017). *Housing for Older People in Singapore: An Annotated Bibliography.* **Cham: Springer**.

This book presented an overview of the policy, practice and research on senior-friendly housing in Singapore. In addition, through desktop literature review of housing for older adults in Singapore, other Asian and developed countries including USA and UK, the annotated bibliography documented the scope of research on housing for older adults. The annotations highlighted the increasing and diverse range of public housing provision for older adults in Singapore while housing finance and quality of life issues were explored from the global perspective.

Yuen, B. (2019). Adapting public housing to age in place in Singapore. In A. P. Lane (Ed.) *Urban environments for healthy ageing: A global perspective.* **London: Routledge**.

This chapter examined the appropriateness and age-friendliness of Singapore's public housing from two perspectives—objective (what's provided) and subjective (how older adults perceive and assess their home environment). Data included the 2013 HDB Sample Household Survey and a 2014–2015 survey with 3025 older Singaporeans aged 55 and above on their experiences of growing old in Singapore. The author also discussed older adults' attitudes towards home modifications and their concerns about future housing needs.

2.2.2   Unpublished Dissertations and Theses

Cheah, W. Y. (1989). *Spatial mobility and social patterns of the aged residents in Queenstown*. **Department of Geography, Faculty of Arts & Social Sciences, National University of Singapore, unpublished undergraduate dissertation**.

The study examined the interaction of 150 older people in Queenstown public housing estate and their surrounding environment during their daily and non-daily spatial mobility. The findings showed that housing characteristics, especially vertical movement between ground and higher floors were critical variables in the study of senior mobility. The housing authority played a significant part in improving senior residents' mobility capability through physical designs of the housing estates. The advocacy is for the transport ministry, grassroots organisations and healthcare institutions to jointly provide more accessible and friendly environment to the older population as well as facilitate their social interactions.

Tan, S. K. (1990). *Life space of the aged in Singapore*. **Department of Geography, Faculty of Arts & Social Sciences, National University of Singapore, unpublished undergraduate dissertation**.

The study looked into the relationship between the personal characteristics of older people in Singapore and their life space as well as spatial mobility. The major personal characteristics such as age, sex, ethnicity, employment status, physiological conditions, and socioeconomic status, all had influence on the patterns of trip activity of older adults. The author also examined the impact of environment on senior mobility. The results indicated that the semi-ambulant and the older frail were the most easily affected groups. It was suggested that the society could enhance senior mobility through age-friendly transportation system and environmental planning.

Sharif Mohamed, S. B. (1993). *Growth of the ageing population: Implications for public housing policies in Singapore*. **Department of Political Science, Faculty of Arts & Social Sciences, National University of Singapore, unpublished undergraduate dissertation**.

This study examined Singapore's public housing policy with respect to older people's housing needs and demands. Current housing policies were analysed and suggestions for modifications were provided with the aim towards an introduction of a senior housing policy, catered towards the future ageing population. Four case studies on the perceptions of older adults' preferred living arrangements in 2030 were presented. Results showed that future senior preferred to live independently within an age-integrated estate, rather than living in a senior-dominated environment. Recommendations for infrastructure and design for old and new estates were also presented in the thesis.

Chan, D. S. F. (1994). *Negotiating space for elderly persons in HDB estates.* **Department of Geography, Faculty of Arts and Social Sciences, National University of Singapore, unpublished undergraduate dissertation.**

This study examined the role of the Singapore government in the allocation of space for older people and the extent to which their needs for living within the community are met. Through interviews conducted with older persons, state authorities and service providers, the study showed that the provision of community-based services was important in meeting the physical and social needs of older residents. However, there was a need to be less top-down by the state, and for more representation by the older people themselves to ensure that their needs were met.

Twoon, W. Y. H. (2018). *Beyond four walls: Recasting informal public spaces as landscapes of care for older persons in Tiong Bahru Orchid.* **Faculty of Arts and Social Sciences, National University of Singapore, unpublished undergraduate dissertation.**

The author argued that informal public spaces in neighbourhoods played a significant part in the care for older adults. It was observed that seniors interacted with one another and caring relations instinctively appeared. The study was conducted on informal places in Tiong Bahru including observations of older residents' face-to-face interactions within these spaces. The author further pointed out that age-friendly designs of informal public spaces could stimulate the older person's experiences of care.

2.2.3   Plans, Reports and Speeches

Fong, C. Y. (1983). Total approach to ageing. *Singapore Elders, 6*(2). **Singapore: Singapore Action Group of Elders.**

The Vice-President of the Singapore Action Group of Elders, Mr. Fong Chan Yoon proposed a 'Total Approach' future plan that was aimed at transforming Singapore into a warm and caring society. The master plan consisted of 18 strategies including encouraging 'three-tier family', setting up more age-friendly public facilities, activity centres and daycare centres, and organising more void deck activities that older people enjoyed.

Inter-Ministerial Committee (IMC). (1999). *Report on the ageing population.* **Singapore: Ministry of Social and Family Development.**

This report outlined the trends, challenges and opportunities faced by the ageing population in Singapore. It contained the policies, plans, services and infrastructure to achieve the vision of Successful Ageing for Singapore. The proposals included social integration of older population, healthcare, financial security, employment and employability, housing and land use policies, cohesion and conflict in an ageing society. One of the key strategies included making our homes and environment senior-friendly and having an integrated community plan to support senior citizens.

Ow, C. H. (2000). *The launch of the safe home programme.* **Presented at Tiong Bahru Community Centre, Singapore.**

Dr. Ow Chin Hock, Minister of State for Foreign Affairs and Mayor for Tanjong Pagar Community Development Council spoke about the Safe Home Programme. Operated by the Tanjong Pagar Community Development Council, the Singapore General Hospital and the Institute of Technical Education, Balestier, the programme was aimed at building an age-friendly living environment that ensured safe homes for older residents.

Committee on Ageing Issues. (2006). *Committee on ageing issues: Report on the ageing population.* **Singapore: Ministry of Community Development, Youth and Sports.**

This report reflected on past efforts and recommended a strategic plan on ageing, detailing how Singapore could be prepared to be senior-friendly in four areas: housing, accessibility, health care and opportunities in maintaining active lifestyles. One desired outcome of the recommendations was to create an age-friendly environment with appropriate housing infrastructure and programmes to support successful ageing.

Tsao, M. A. (2013). Mapping out an age friendly Singapore: Lessons from pioneering work in ageing and eldercare. In *Social space* **(pp. 4–9). Singapore: Lien Centre for Social Innovation at Singapore Management University.**

This interview with Dr. Mary Ann Tsao, Chairman, Tsao Foundation, focused on Tsao Foundation's achievements and initiatives on ageing-related issues, policies, and programmes. It recorded her perspectives on the challenges faced as well as the impact made by the Foundation, and future involvement in creating an age-friendly neighbourhood.

Goh, L. P. (2014). *Public housing in Singapore: Social well-being of HDB communities.* **Singapore: Housing & Development Board.**

This reported the residential satisfaction survey with public housing residents, the data and findings relating to senior-friendly housing features and neighbourhood provisions such as grab bars, wheelchair accessibility, availability of eldercare centres and equipment.

Ministry of Health. (2014). *Creating senior-friendly communities: Tips and tools from the city for all ages project.* **Singapore: Ministry of Health.**

This guidebook presented the collective experiences and ideas from various communities in the City for All Ages Programme, a nationwide initiative to make Singapore senior-friendly. It outlined how an individual could start or be part of this initiative and the possible funding sources. Tools and current efforts to make neighbourhoods senior-friendly such as town audits were detailed.

Permanent Mission of Singapore. (2015). *The interactive dialogue with special rapporteur on the rights of older persons***. Presented at the 30th Session of Human Rights Council, Switzerland**.

The Permanent Mission of Singapore shared Singapore's plan to meet the challenges of an ageing population. This included providing the ageing population with a better living environment through the development of more nursing homes, renovation of public infrastructure and establishment of age-friendly policies and programmes. A key strategy was Singapore's urban planning, which ensured that residential environments and the city were senior-friendly.

Ministry of Health. (2016). *Action plan for successful ageing***. Singapore: Ministry of Health**.

This report detailed an action plan for successful ageing based on the ideas and discussions with the community, public and private sectors. Ten topics were presented and discussed with suggested plans for a positive outcome. The topics included keeping good mental and physical health, being part of society, contributing back to society, developing senior-friendly public spaces and housing that would meet ageing needs, providing services for older adults, continual learning and mobility.

Housing & Development Board. (2017). *Good neighbours, great communities***. Singapore: Housing & Development Board**.

This publication featured examples of how individuals and groups played a part in their neighbourhoods and contributed to building a great community. One story shared how Madam Veerama d/o Thalaniyandi Sivalinga visited her older neighbours weekly to get to know them and their demands, offered them help and resources through the WeCare@MarineParade Programme. Another example showed the efforts of Welcome to Our Backyard (WOBY!) project, in which residents at Aljunied Crescent worked closely with grassroots leaders, older residents and the Geylang East Home for the Aged to transform an underused space in the neighbourhood into a vibrant activity zone for both young and old.

Loo, D. (2017). *Successful ageing in Singapore: Urban implications in a high-density city***. Singapore: Lee Kuan Yew School of Public Policy, National University of Singapore**.

This paper reviewed the policy related to ageing and urban development in Singapore. It provided details of ageing policy developments, events and projects since the 1980's. In particular, it detailed the efforts and plans for ageing-friendly city, housing, neighbourhoods, facilities, services and environment for older people to live active lives and participate in community activities.

National Council of Social Service. (2017). *Understanding the quality of life of seniors.* **Singapore: National Council of Social Service.**

The report began with various relevant statistics on demographic ageing in Singapore such as increased life expectancy, healthy years, rising numbers of older adults and age-related conditions. It argued that stereotypes about old age often represented negative associations such as unproductive, boring, incapable, and should be discarded. Survey findings on what had contributed to a higher quality of life and current efforts undertaken to make neighbourhoods age-friendly were highlighted.

Lee, H. L. (2018). *PM Lee Hsien Loong at the Opening of Kampung Admiralty.* **Presented at Kampung Admiralty, Singapore.**

This speech was delivered by the Prime Minister at the opening of the Kampung Admiralty public housing project. This marked the introduction of a new public housing typology on integrated senior-friendly neighbourhood, featuring a medical centre, active ageing hub and a community plaza where 50 senior activities had been organized for the day.

Khor, A. (2019). *Speech by Dr Amy Khor, Senior Minister of State for Health, in response to motions on Ageing with Purpose and Support for Caregivers.* **Presented at Parliament House, Singapore.**

This speech provided an update on the progress and achievements of active ageing initiatives relating to healthcare provision and promotion as well as future plans to enhance ageing in place programmes. It mentioned new schemes and activities to further support senior inclusion in neighbourhoods.

2.2.4 Media Reports

2.2.4.1 Community Centres, Programmes and Activities

(1980, August 20). 'Meals on wheels' alternative being examined. *New Nation, Singapore***, p. 6.**

The Singapore Action Group of Elders (SAGE) proposed to issue allowance to the neighbours or nearby food stalls of bed-ridden older people, if the neighbourhood was willing to cook meals or deliver hawker food them. Dr. Lim Chan Yong, President of SAGE, suggested this as an alternative to the "Meals on Wheels" Programme.

(1983, May 27). A club for the aged—Run by the aged. *The Straits Times, Singapore***, p. 10.**

The People's Association Retirees Club looked forward to getting senior citizens to look after the club, playing both the roles of leaders and members. It planned to help older people be aware of their usefulness in the society as well as their potential contributions to the community.

(1983, July 25). Bringing hobby courses to elderly. *The Straits Times, Singapore,* **p. 2**.

Three organisations—the Singapore Women's Association, National University of Singapore Department of Extramural Studies and Singapore Actions Group of Elders had joined hands to organise a broad variety of interest courses for older people. The aim was to enable older people to maintain independence and confidence as well as enjoy a meaningful life.

(1983, August 19). Aged and disabled 'have a role on CC committees'. *The Straits Times, Singapore,* **p. 17**.

Mr. Teo Chong Tee, the Parliamentary Secretary and Member of Parliament for Changi, said that older people and handicapped people should be supported to take their place in community centres committees and sub-committees. This would give them more opportunities to get involved and contribute to community activities.

(1983, August 28). Club where elderly will have chance to help the elderly. *The Straits Times, Singapore,* **p. 7**.

The People's Association Retirees Club initiated a wide range of programmes, encouraging its older members to help out other senior residents in the neighbourhood including a keep-in-touch scheme and an elders' friendship programme, in which senior volunteers visited senior residents, kept them company and took care of them.

(1984, April 18). PA to organise more activities for increasing number of English-educated senior citizens. *Singapore Monitor, Singapore*.

The People's Association would put more efforts into organising activities for the new generation of senior citizens who were increasingly English-educated and had new and different demands and interests from past cohorts. Other than the activities specially designed for seniors, the association was looking at bringing together younger and older people.

(1984, June 15). PA to set up senior citizens' section. *The Straits Times, Singapore,* **p. 8**.

The People's Association would launch a senior citizens' section to organise activities for older people and work on senior-related projects such as health screening and talentime. The goal was to encourage the seniors to stay active and connected in the community.

(1984, June 15). Senior citizens should keep active even after retirement. *Business Times, Singapore,* **p. 2**.

Mrs. Teresa Tsien, Assistant Director at the People's Association, recommended several activities to keep seniors active while ageing, namely, educational courses, recreational tours and cultural pursuits. She also suggested that older people should participate in fitness programmes to stay healthy. In addition, making contributions to the society through volunteer work and befrienders' programmes could also help the older person stay active.

(1984, August 7). When help is just a neighbour away. *The Straits Times, Singapore,* **p. 14**.

The Singapore Action Group of Elders launched a pioneer programme, encouraging the neighbourhoods in Maude and Kitchener Roads to look after each other and older people within the community. This would help to reduce the stress borne by the families of older people, especially those who were chronically ill or bedridden.

(1984, August 15). Focus on senior citizens. *The Straits Times, Singapore,* **p. 60**.

In a programme called 'The Citizen', three senior citizens shared their feelings and viewpoints of being an employed older person, a retiree and a would-be retiree. The latter half of the programme focused on how the senior citizens' clubs supported older people and helped them to keep fit and active through age-friendly activities.

(1985, January 29). Friendship beyond old folk's homes. *The Straits Times, Singapore*, **p. 11**.

The Henderson senior citizens' home corner, run by the Singapore Council of Social Service, had become a venue where older people from all over Singapore gathered regularly, catching up with one another, playing games and having meals together. The Council encouraged other senior citizens' housing to set up such age-friendly resident corners.

(1986, May 16). Club that's like a second home to 200 old folk. *The Straits Times, Singapore*, **p. 21**.

The Havelock community centre served as a second home to its 200 members, many of whom were older residents. The Centre not only offered a space where the old residents could meet and chat but also organised age-friendly games and activities.

(1986, September 23). Activities keep old folk busy. *The Straits Times, Singapore*, **p. 12**.

The senior citizens' clubs set up by different community organisations in the neighbourhoods provided a broad variety of activities to the ageing population so as to keep them active in the communities. In addition, various programmes such as daycare and healthcare centres, fitness schemes and befrienders initiatives were launched by the Government for "helping senior citizens to help themselves".

(1987, July 14). A morning walk and then qigong for the elderly. *The Straits Times, Singapore*, **p. 15**.

Some 4000 old folks from 65 senior citizens' clubs and the People's Association Retirees Club participated in a walkathon-cum-qigong activity, which was one of the many programmes initiated and promoted by the Association to encourage older people to stay healthy and active.

(1988, February 19). The qigong breakfast club. *The Straits Times, Singapore*, **p. 18**.

Qigong enthusiasts, comprising senior and middle-aged citizens, organised daily breakfast gatherings at Bedok North neighbourhood after their exercise, during which the members socialised with each other. This new way of getting together allowed the older person to become more active in the community.

(1992, June 6). Dumpling treat for the elderly. *The Straits Times, Singapore*, **p. 29**.

The Marine Parade Community Club held a rice dumpling treat for 120 older people from five homes for the aged during the Dragon Boat Festival. The aim was to recreate the festive feeling for these residents.

Tan, C. (1992, October 9). Group on a friendly mission to bring cheer to the elderly. *The Straits Times, Singapore*, **p. 33**.

The Centre of Activity and Care for Elders, formed by a group of middle-age businessmen and community leaders, organised recreational activities for senior citizens in order to enrich their lives. In addition, the Centre offered both financial and medical support to needy older people including continuous learning and healthcare.

Cheng, C. S. (1992, November 15). Happy, healthy and elderly. *The Straits Times, Singapore*, **p. SR14**.

The article suggested that older people could engage themselves in a wide range of activities so as to keep active while ageing such as doing exercise, gardening and participating in club activities. It reviewed the courses and activities offered by several organisations including the People's Association, Singapore Action Group of Elders and Senior Citizen Clubs by Resident Committees.

(1993, July 9). Free lunches for the elderly poor when centre opens. *The Straits Times, Singapore*, **p. 31**.

The MacPherson Moral Family Service Centre provided free lunch meals to seniors of all races under the Government Public Assistance Scheme. In addition, the Centre offered services and facilities such as daycare and healthcare assistance, computer lessons, a library and family consultation for older people.

(1993, September 15). Leisure for elderly. *The Straits Times, Singapore*, **p. L2**.

The article reviewed the leisure opportunities for older Singaporeans. Some private senior citizens' centres, for instance, the Evergreen Cultural and Recreational Centre, provided a wide variety of activities and lessons for retirees such as calligraphy and handicraft. Government-owned eldercare centres offered healthcare and fitness programmes for older people, e.g. the bi-annual Middle-Aged and Elderly Athletic Meet.

(1993, October 5). Neighbourhood corner for the elderly offers free meals. *The Straits Times, Singapore*, **p. 25**.

The Bishan-Serangoon Town Council had established a neighbourhood corner at its void deck area with age-friendly features such as railings, tables and benches. The neighbourhood corner not only provided a relaxing environment for older residents but also a venue where the community supplied free meals to the needy.

(1995, November 30). Elderly volunteer helps other aged neighbours. *The Straits Times, Singapore*, **p. 37**.

Located at the void deck of a MacPherson neighbourhood, the Moral Seniors Activity Centre was among the first piloted Seniors Activity Centres by the government. The major purpose of setting up this centre was to take care of the solitary older people in the neighbourhood. It was equipped with exercise machines and karaoke equipment. All flats in the apartment block were connected to the centre by a computerised alarm system.

(1996, January 26). Three-in-one activities at Tampines East CC. *The Straits Times, Singapore*, **p. L22**.

An activity centre at the Tampines East Community Club had adopted a 'three-in-one' service concept by integrating the services for senior citizens, pre-school and primary school children. The approach encouraged the senior resident to remain active within the community and interact with the youth and children.

Wong, C. M. (1996, December 13). Young cooking up a storm for elderly. *The Straits Times, Singapore*, **p. L30**.

Youth volunteers in their 20's cooked and prepared meals for over 70 older people at a senior citizens' club in Hong Lim. As a volunteer commented, she enjoyed a lot through her interactions with older people.

(1999, June 12). Youths do their part to help the needy. *The Straits Times, Singapore*, **p. 54**.

The Northeast Community Development Centre initiated a programme, Youths in Community Action, involving 3000 students aged between 15 and 19 years to spend some time during their school years to participate in nationwide community activities. These community activities included providing lunch meals for older people and organising games and performances for needy families.

(1999, August 21). Aged get activity centre. *The Straits Times, Singapore*, **p. 42**.

The Touch Seniors Activity Centre offered close-to-home medical and community care to 150 living alone senior residents in Geylang Bahru. The centre aimed to provide these residents with a sense of belonging and help them to make friends through activities.

Yen, F. (2005, November 12). Befrienders like family to lonely elderly. *The Straits Times, Singapore,* **p. H16.**

Lions Befriender volunteers visited and accompanied lonely older people in the community regularly, talking to them and helping them with groceries and phone bills. With their care and support, these older residents had become more active and developed better physical and mental status.

Yap, S. Y. (2006, March 10). Keeping senior citizens on the Go with $20 m fund. *The Straits Times, Singapore,* **p. H5.**

Over 200 senior citizens had actively registered for various courses provided by the Yah! Community College at the Marine Parade Family Service Centre. More life-enrichment programmes, aimed at maintaining the physical and mental status of older people, were in the pipeline with substantial financial support from the government.

Sim, M. (2010, March 13). More seniors now socially active: HDB. *The Straits Times, Singapore,* **p. B4.**

In a survey of 865 older households, it was reported that older residents living in Housing and Development Board flats had actively participated in community activities organised by community clubs, residents' committees and religious organisations. Through these activities, they made new friends, reunited with old friends and enjoyed a higher quality of life.

Chin, D. (2012, December 17). Hobby groups help seniors stay active, make friends. *The Straits Times, Singapore,* **p. B7.**

Interest groups such as ukulele, qigong and balloon sculpting, enabled older residents to connect with their peers of common interests and made friends in the neighbourhood. The sessions had been well received amongst the older population due to their easy entry.

Yeo, D. (2013, January 7). Making Singapore a city for all ages. *Public Service Division, Singapore.*

This article highlighted how the bottom-up initiative, City for All Ages from Marine Parade had developed into a nationwide programme. The initiative focused on identifying service gaps in the neighbourhood that could help older people to age gracefully through a series of surveys, health checks and town audits.

Mehta, K. (2013, August 16). Creating an age-friendly Singapore requires more care. *TODAYonline, Singapore.*

Comparing Singapore's efforts with international guidelines, the article argued that besides physical infrastructure, an inclusive, senior-friendly neighbourhood would require social development. In particular, older adults' involvement in society was often overlooked, especially when involving older people in the conceptualisation and execution of programmes and initiatives for older people.

Yong, C. (2013, August 12). Ayer Rajah breakfast club helps elderly find friends. *The Straits Times, Singapore*, p. B3.

At the Ayer Rajah public housing estate, free breakfast and various activities were provided to senior citizens. It provided opportunities for older people in the neighbourhood to socialise with the community and meet new friends instead of staying disconnected and isolated at home.

Tai, J. (2016, January 29). Seniors cook up camaraderie in novel void deck kitchen. *The Straits Times, Singapore*, p. B1.

In the newly initiated open-concept kitchen project, senior residents at Marine Parade were given the opportunity of cooking their own meals within the neighbourhood. This allowed them to not only participate in the activities but also contribute to the community. The project also connected older people and removed the social identity of 'stay-alone seniors' from them.

Au-Yong, R. (2016, July 29). People's Association to offer more courses to help senior citizens stay active. *The Straits Times Online, Singapore*.

The People's Association issued an Advanced Certificate in Senior Wellness to support continuous learning amongst older people, e.g. to learn a musical instrument, cooking and crafts skills. This would help and encourage older people to stay active as they age.

Au-Yong, R. (2016, November 3). Bowling seniors over with Kinect technology. *The Straits Times Online, Singapore*.

At the senior activity centre in Toa Payoh, senior residents embraced Kinect technology through playing bowling. According to one older adult, the game was easy to pick up, trained them to stay focused and kept them exercising as well.

Chan, F. (2017, July 3). Make your neighbourhood a better place. *The Straits Times, Singapore*, p. 5.

The Housing and Development Board launched various volunteer programmes under its Friends of Our Heartlands Network with the aim to connect people with their communities. For instance, as part of the Activate track of activities, volunteers were encouraged to engage and communicate with older people, and make the housing surrounding more cheerful.

Yong, C. (2017, July 24). Pioneer Generation envoys doing more to help seniors. *The Straits Times, Singapore*, p. A6.

Since the Pioneer Generation Ambassador Programme was launched in 2014, an increasing number of ambassadors had helped seniors to age well in their communities. It was announced by Prime Minister, Mr. Lee Hsien Loong that the programme would spend more efforts in motivating older people to exercise and make friends within the neighbourhood.

Tan, T. M. (2017, October 2). More Silver Academy courses for seniors. *The Straits Times, Singapore*, **p. B3**.

The National Silver Academy aimed to launch over 900 new courses in 2017 and offer 21,000 more learning places to encourage senior citizens to lead active lives with a variety of subjects. Dr. Amy Khor, Senior Minister of State for Health, announced that the efforts had doubled in 2017 as compared to 2016.

Abdullah, Z. (2017, November 12). Active ageing hub opens in McNair Road. *The Straits Times, Singapore*, **p. A9**.

As part of the Ministry of Health plan to develop 10 active ageing hubs within new public housing areas, Kwong Wai Shiu Hospital launched its first care centre at McNair Road to serve seniors living in the Kallang and Whampoa region. Rehabilitative care and social activities were provided at the centre with an aim to encourage seniors to stay active while ageing.

Tan, S. Y. (2018, May 5). New training facility to better equip befrienders of the elderly. *The Straits Times, Singapore*, **p. B9**.

Under the new Community Networks for Seniors programme, 1700 SG Ambassadors had gone through training to gain skills in engaging and befriending with the ageing population as well as providing health care and promoting active ageing among the older population. The venue for this training would be a new facility at the Silver Generation Office in the Ministry of National Development.

Teoh, Z. W., & Zainal, K. (2018, July 6). Successful ageing: Progressive governance and collaborative communities. *Civil Service College, Singapore*.

With a focus on the Factsheet on Action Plan for Successful Ageing, this article highlighted the need for close collaboration between different agencies, businesses and sectors in order to transform Singapore's housing estates into senior-friendly neighbourhoods.

Rashith, R. (2019, February 23). Budget also aims to 'help seniors remain active'. *The Straits Times, Singapore*, **p. B4**.

At the opening of an age-friendly gym run by the voluntary welfare organisation, Montfort Care, Second Minister for Finance and Education, Ms. Indranee Rajah remarked that healthcare budget for older people and senior activity centres in the community both played critical roles in ageing support while senior volunteerism offered another helpful approach to active ageing.

Wong, C. (2019, March 9). Parliament: More centres to mobilise volunteers in neighbourhoods. *The Straits Times Online, Singapore*.

Ms. Grace Fu, Minister for Culture, Community and Youth, shared that two Volunteer Centres had been set up in Bedok and Jurong East as hubs for volunteer recruitment and development before they embark on volunteer programmes in the communities. The Filos Community Services in Bedok would focus on volunteer engagement and

Jurong's Loving Heart Multi-Service Centre would work on a befriending Neighbour Cares Programme. Five more such volunteers centres were planned.

Yip, W. Y. (2019, May 27). Temasek Poly links up with HDB to boost active ageing activities. *The Straits Times, Singapore*, p. B6.

Temasek Polytechnic started a three-year partnership with the Housing and Development Board, in which students reading gerontology would reach out to seniors in their neighbourhoods through activities such as fitness exercises and crafts workshops. The goal was to help older residents to live healthy and active lives. The programme began with neighbourhoods in the east and the north-eastern part of Singapore including Punggol, Sengkang, Hougang, Bedok, Tampines and Pasir Ris.

2.2.4.2 Eldercare Services and Support Infrastructure

(1980, February 18). Old and lonely need social contacts. *The Straits Times, Singapore*, p. 5.

The Singapore Council of Social Service did a survey with 535 older people to identify senior citizens' demand for a daycare centre and found that 85% of respondents were in need of companionship and social-recreational activities. The conclusion was that a daycare centre was necessary and significant in the ageing society.

(1981, June 21). Model centre for elderly in HDB estates opens. *The Straits Times, Singapore*, p. 9.

The therapeutic day care centre in Bukit Merah was opened by the Singapore Council of Social Service to cater to the ageing population as well as their families. It was designed as a model that other community groups could learn from to establish similar eldercare services and facilities in their respective neighbourhoods. The Acting Minister for Social Affairs, Dr. Ahmad Mattar encouraged voluntary organisations to provide more care and support to older people.

(1982, April 6). Sage elders' village to be built in Toa Payoh. *The Straits Times, Singapore*, p. 40.

The Singapore Action Group of Elders (SAGE) launched a new Elders' Village complex at Toa Payoh West, which consisted of 300 units that could each accommodate a whole family. The complex was equipped with recreational and medical facilities. SAGE expected the whole building to be managed by older people as a way to help them stay active and contribute to the community.

(1982, April 12). Make life easier for the aged call to govt and community. *The Straits Times, Singapore*, p. 1.

The Singapore Council of Social Service in its recent report indicated that older people were in greater need of healthcare, housing, financial and employment support.

It further argued that social service and welfare organisations should launch more daycare centres and senior citizens' clubs in the neighbourhoods as well as organise more age-friendly activities.

(1982, July 7). Planning to give senior citizens a new deal. *The Straits Times, Singapore,* **p. 13.**

The Kampong Kembangan senior citizens group proposed some suggestions to the Ministry of Health on how to make older residents' lives better in the community. These included enabling older adults to be ready for retirement, providing more daycare centres for older population, and launching more senior citizens' centres at the void deck of public housing blocks.

(1983, February 27). Knock, knock it's your friendly policeman. *The Straits Times, Singapore,* **p. 1.**

The Neighbourhood Police Posts embedded within the communities served an important role in maintaining the safety of the neighbourhoods and could be called upon to help look in on the older population. For example, when young people were at work during the daytime, they could soon be alerted by the police if their older parents fell sick at home.

(1984, July 27). New haven for the aged folk. *The Straits Times, Singapore,* **p. 13.**

The Ling Kwang Home for Senior Citizens had become a warm home for 116 old folks who were provided with a wide range of programmes at the Home such as spiritual counselling, exercise classes, therapy services, etc. The Home was frequently visited by youths, allowing the older residents opportunity to maintain good social contact with people of other ages.

(1984, October 26). Young bring joy to the aged at kongsi. *The Straits Times, Singapore,* **p. 17.**

The Rochore Kongsi was set up to cater to the older residents in the neighbourhood with financial support from both the neighbours and nearby shop owners. The Kongsi was equipped with 4 hostels, 8 toilets, 4 bathrooms and a void deck where older residents enjoyed the company of children from the neighbourhood.

(1984, December 20). Plan to give elderly better health care. *Business Times, Singapore,* **p. 2.**

The Home Nursing Foundation planned to launch a more holistic healthcare scheme for senior citizens including establishing 6 rehabilitation centres in public housing areas, providing healthcare courses and counselling services, and offering medical check-ups. The programme was largely supported by the Khoo Foundation. It could subsequently involve more health professionals and deploy more medical facilities to cater for the ageing population.

(1985, January 19). Community's role in care for the aged. *Business Times, Singapore*, **p. 2**.

Parliamentary Secretary, Mr. Wan Hussin Zoohri pointed out that with an increasing number of older people being left alone at home, external agencies, especially voluntary organisations, could provide home-based care and support to older people within the neighbourhoods.

(1985, May 8). Health care service at your doorstep. *The Straits Times, Singapore*, **p. 21**.

The Home Nursing Foundation planned to launch 10 community health centres in public housing areas to make rehabilitative and healthcare services available at the doorstep of the ageing population, especially those with mobility issues. The initiative was aimed at enabling older people to live independently, keeping them active, and encouraging the semi-mobile older adults not to become bedridden. The Foundation announced its master plan for ageing society with focus on four major areas—identifying older people's needs, engaging them in the community, collaborating with community organisations, and promoting the Foundation's programmes.

(1986, January 25). Health centres for the aged. *The Straits Times, Singapore*, **p. 1**.

The government announced plans to set up 2 healthcare centres for the ageing population on an annual basis. The aim was to enable older adults to stay within their own community with the help and support from their families, friends and neighbours rather than being institutionalised. The plan included providing healthcare services to seniors at community levels and establishing hospitals in the neighbourhoods to attend to the frail seniors.

(1986, May 16). More facilities and services are needed. *The Straits Times, Singapore*, **p. 21**.

Two university professors, Dr. S. Vasoo and Dr. Aline Wong, commented that the current social services and facilities provided in the communities were insufficient to meet the needs of the ageing population. For example, the coverage of the services and facilities should be expanded to serve a wider population, some clubs should offer appropriate activities to engage the seniors, and more daycare centres should be deployed in the neighbourhoods so that the older population could grow old within the communities where they used to stay.

(1986, September 27). Schemes to make life a bit easier for the old folk. *The Straits Times, Singapore*, **p. 13**.

The Home-Help Service initiated by the Ministry of Community Development aimed to enable older people to lead a better life as well as relieve their families from the heavy load of nursing work. The programme included a recruitment drive to employ more home helpers, a Befriender Service to encourage volunteers to attend to those in need, and a Respite Service for caregivers to allow seniors to receive care and support at voluntary homes.

Low, A. (1986, October 13). Free eye and ear checks for elderly. *The Straits Times, Singapore*, **p. 11**.

The Home Nursing Foundation provided eye and ear examinations at no cost to senior citizens at the Senior Citizens Health Care Centre in Toa Payoh. The Foundation planned to build 10 more healthcare centres in public housing areas by 1990, aimed at enabling older people to live more independently in the communities.

(1989, July 27). Helping the populace to grow old gracefully. *The Straits Times, Singapore*, **p. 3**.

Six more senior citizens healthcare centres were to be set up by the Home Nursing Foundation in public residential areas by 1992. The goal was to allow older people to stay within the communities rather than living in nursing homes through various community-based eldercare services.

(1991, March 1). Home-based services needed for the sick and elderly. *The Straits Times, Singapore*, **p. 28**.

Nominated Member of Parliament, Mr. Maurice Choo commented that a broad range of home-based services should be provided to the ageing population including neighbourhood clinics, daycare centres, family doctors and home nursing. In addition, older people should be encouraged to join the workforce, to enable them to financially support themselves, and make significant contributions to the society.

(1991, March 3). Day-care centre for Hougang's elderly. *The Straits Times, Singapore*, **p. 15**.

The Hougang Polyclinic hosted a Senior Citizens Health Care Centre where older adults were able to rest, exercise, catch up with their peers, and enjoy physiotherapy and counselling services. The centre also recruited professional caregivers to look after these older people and train family members to take care of their older members.

(1991, June 4). Drop-in day centre for elderly. *The Straits Times, Singapore*, **p. 21**.

Senior Activity Centres provided older people with a wide variety of activities for recreation and education at low or no cost including lessons on handicraft skills, foreign language courses and sports games. The programmes enabled older people to be actively involved in the community and interact with their peers.

(1993, September 23). Health care for elderly at doorstep. *The New Paper, Singapore*, **p. 5**.

The Ministry of Health and Community Development announced that more daycare centres would be built in housing areas so that older people could stay in the comfort of their familiar communities while visiting the healthcare centres instead of being warded in hospitals.

(1995, July 18). Visionary club for aged. *The New Paper, Singapore*, **p. 4**.

It was reported that Pulau Ubin would be transformed into a new town with first-class amenities, a senior citizens' club and facilities tailored for older population. Some 29 Temasek Polytechnic architecture students were involved in a design-prototype project for the senior citizens club, which was led by the National Council of Social Service.

(1997, August 12). Have care centres for young and old 'near each other'. *The Straits Times, Singapore*, **p. 31**.

NTUC Chief, Mr. Lim Boon Heng advocated that in order to strengthen and preserve family ties, eldercare centres should be set up near to the childcare centres so that the middle generation would be able to take care of both their children and parents at the same time.

(1998, April 17). New day-care centre for elderly. *The Straits Times, Singapore*, **p. 3.**

The NTUC Eldercare initiated a fund-raising programme for a new daycare centre in a neighbourhood that would also provide affordable meals, various activities and health care services to older people. The cooperative also looked at collaborating with other organisations to provide age-friendly services such as caregiver training programme and pre-retirement lectures.

(1998, April 20). Care for elderly. *The New Paper, Singapore*, **p. 4**.

Older residents living at Marine Parade could borrow walking sticks, wheelchairs and hospital beds from a care centre located in Sims Avenue. The centre also provided other age-friendly services such as health screenings, home safety checks, home therapy and nursing.

Chin, S.F. (1999, March 26). Help for the elderly under one roof. *The Straits Times, Singapore*, **p. 54**.

The Asian Women's Welfare Association provided affordable, convenient one-stop services to older people living in Ang Mo Kio through co-locating a daycare centre and an activity centre side-by-side. The centres offered physiotherapy services, cooking and handicraft activities, healthy meals, and more.

(1999, May 28). Library sets up room for elderly. *The Straits Times, Singapore*, **p. 75**.

The Toa Payoh Community Library opened Singapore's first senior citizens' room with around 200 special large print books, sofas, chairs and a multimedia computer. The National Library Board aimed to encourage older people to continue reading and learning as they aged as well as pick up some information technology skills.

Chin, S. F. (1999, May 31). Centre for the elderly to open. *The Straits Times, Singapore*, **p. 23**.

The Community Action for the Rehabilitation for the Elderly (CARE) Centre provided rehabilitative care, daycare and health screenings to frail seniors living in the Northeast Community Development Council district so as to enable older people to live independently in the neighbourhood.

Boo, K. (1999, August 27). One-stop care for needy elderly. *The Straits Times, Singapore*, **p. 34**.

The Pasir Ris Nursing Home was a pioneer in offering multiple eldercare services to the ageing population. The services included polyclinic, nursing home, counselling services, minimart and loanable equipment such as walking aids and wheelchairs.

(1999, September 4). Elderly-care pilot project expands. *The Straits Times, Singapore*, **p. 64**.

The National Council of Social Service launched the Case Management Service Programme that provided a comprehensive healthcare scheme to the ageing population by assigning case managers to senior citizens who were discharged from hospitals. The case manager would identify the medical and nursing needs of older people, and then deal with care providing agencies to deliver customised services to them.

Khalik, S. (2000, August 24). Comprehensive health services for the elderly. *The Straits Times, Singapore*, **p. 36**.

The Ministry of Health was offering a wide range of healthcare services for older people including community health screening, family doctors, continuing care for the aged, etc. Through these programmes, the government endeavoured to raise the awareness of keeping fit and taking necessary treatment amongst the older population.

(2000, November 19). Centre to care for elderly to open in Yishun. *The Straits Times, Singapore*, **p. 38**.

To cope with the increasing needs of the ageing population in the neighbourhood, the newly established Yishun Eldercare Centre offered a place where older people could take part in leisure activities and enjoy the various healthcare services such as physical maintenance and rehabilitation.

Liew, H. (2001, January 13). $93 m masterplan for care of the elderly. *The Business Times, Singapore*, **p. 7**.

The Eldercare Master Plan was approved by the Ministry of Community Development and Sports. Among its many provisions, it highlighted that Multi-Service Centres and Neighbourhood Links would play a significant role in providing care and support to older people at the community and neighbourhood levels. In addition, the government planned to subsidise eldercare services on a per capita basis.

(2006, November 21). Mobile health-care service for the elderly. *The Straits Times, Singapore*, **p. H2**.

Free and affordable healthcare services on a bus would soon be available for senior residents in 10 housing areas. The mobile healthcare services would provide health screenings and traditional Chinese medicine at various public housing neighbourhoods including Hougang, Sengkang, Punggol.

Hussain, Z. (2006, November 24). Best way to age? At home with help from neighbours. *The Straits Times, Singapore*, **p. H2**.

More 'home help' services were needed to enable older residents to live independently within a community. These included building up a community-based support system as well as a neighbourhood assistance network, in which older people would be attended to by people living nearby whenever they were in need.

Yap, S. Y. (2006, December 14). Home sweet retirement home. *The Straits Times, Singapore*, **p. H8**.

At Jalan Jurong Kechil, senior citizens were expecting a developer to turn a residential site into a friendly and affordable retirement village. The older people were looking forward to retirement units with barrier-free facilities, elder-friendly amenities and nursing services.

Jiang, G. (2008, October 22). Help can be found right at seniors' doorstep. *The New Paper, Singapore*, **p. 7**.

The Ministry of Community Development, Youth and Sports had set up around 40 senior activity centres and neighbourhood links across Singapore. The aim was to encourage seniors to get together for exercise, lessons, games and activities. To prevent older adults from ageing alone, the Government also called on neighbourhood residents to help look after their older neighbours.

(2009, February 6). More day care, activity centres for the elderly. *The Straits Times, Singapore*, **p. C15**.

It was reported that the government would launch 6 new daycare centres and 22 senior activity centres within five years, aiming at providing better care and support to the older population. New training and support programmes for caregivers would also be introduced.

Ong C., Tai, J. (2011, November 17). Help on the doorstep for seniors. *The Straits Times, Singapore*, **p. B4**.

The government collaborated with voluntary groups to launch more healthcare facilities at the void deck of seniors studio apartment block including daycare centres and seniors service centres. The initiative was to mainly cater to older people who were unable to access family members for eldercare.

Ong, C. (2011, December 5). Health checks for seniors at roadshows. *The Straits Times, Singapore*, **p. B3**.

A new series of health screening roadshow, Healthy Living @ South West, was rolled out in public housing neighbourhoods like Jurong and Choa Chu Kang. This programme would enable senior citizens to more easily access health check-ups. Such initiative would cover 17 more divisions in Southwest District within three years.

Basu, R. (2012, February 22). 'Yes, we need retirement villages'. *The Straits Times, Singapore*, **p. A10**.

It was reported that there was a need for retirement villages amongst the ageing population where senior-friendly housing was equipped with holistic on-site services such as health care, community engagement and professional management.

Tay, S. C. (2012, February 29). At group home, seniors find freedom and support. *The Straits Times, Singapore*, **p. B8**.

The Seniors Group Home initiated by the Thye Hua Kwan Moral Society brought together a group of older people who had minimal or no support from their families. Instead of being sent to nursing homes, the Group Home allowed these older people to remain independent within a community and live together while having the freedom to do their own things.

Khalik, S. (2013, May 17). Vital for eldercare centres to be within community. *The Straits Times, Singapore*, **p. A1**.

It was reported that the eldercare infrastructure should be embedded within the community. This would be more accessible for the ageing population to enjoy eldercare services as well as foster a more comprehensive and engaging healthcare environment for older people.

Tai, J. (2014, April 19). More help for lone seniors under new eldercare system. *The Straits Times, Singapore*, **p. A3**.

A new eldercare system was to be established by the Ministry of Social and Family Development across Singapore. Up to five senior activity centres would be set up to accommodate the social and recreational activities organised for older people, and three group homes would be built for the ageing group who had minimal or no family support.

Lien, L. (2014, November 22). Time to help seniors age successfully. *The Straits Times, Singapore*, **p. A40**.

It was important to enable older people to stay in the neighbourhoods where they had lived a number of years. To realise this vision of ageing in place, eldercare services were significant and should be introduced to the ageing communities. This would allow older people to enjoy various services and facilities in familiar environments.

Tan, W. (2016, April 14). What seniors need, care providers seek to meet. *The Straits Times, Singapore,* **p. B4**.

The Ministry of Health collaborated with care providers to launch a trial programme under the Integrated Home and Daycare package that provided customised eldercare services to older people such as nursing homes, eldercare centres and home care.

Tay, T.F. (2017, February 27). Elderly care centre launched at upgraded Whampoa CC. *The Straits Times Online, Singapore*.

The Community of Successful Ageing Centre was opened to provide healthcare, social care and community services to ageing residents at Whampoa. The care infrastructure enabled better transition between hospitals and home, providing comprehensive healthcare advisory and services to older residents.

Chuang, P. M. (2018, February 20). S\$550 m increase in spending on health and social services. *The Business Times, Singapore,* **p. 11**.

Mr. Heng Swee Keat, Minister for Finance, announced in the Budget Speech that the government would merge the health and social services for older people to better serve this population. Another S\$550 million budget would be spent. Programmes that encouraged active ageing and befriending, and provided health and social support were anticipated to reach out to more older people.

Khalik, S. (2018, July 7). Services to meet seniors' long-term care needs. *The Straits Times, Singapore,* **p. B1**.

Thanks to the daycare services provided by the Salvation Army Peacehaven Bedok Multi-Service Centre and subsidised transport offered through the Seniors' Mobility and Enabling Fund by the Agency for Integrated Care, stroke patient, Madam Oei was able to receive care while continuing to stay at home. Mr. Gan Kim Yong, Minister for Health, remarked that more efforts would be put to serve senior citizens like Madam Oei.

Rashith, R., & Khalik, S. (2018, August 15). MOH to grow aged care services to meet rising demand. *The Straits Times, Singapore,* **p. A6**.

In response to a study carried out by Lien Foundation, the Ministry of Health announced that it would keep expanding its services to support Singapore's ageing population. Efforts from the Community Networks for Seniors had already seen weekly active ageing programmes continuous growth in over 360 neighbourhoods across Singapore. In terms of financial aids, qualified senior citizens were able to get up to 80% subsidy from the government.

Lim, M. Z. (2018, September 23). New hub in Tampines to meet seniors' health and social needs. *The Straits Times, Singapore*, p. A8.

The new SilverCare Hub was opened at Our Tampines Hub to provide health and social support to senior residents in the eastern region of Singapore so as to help them stay healthy and active. Mr. Heng Swee Keat, Minister for Finance and Member of Parliament for Tampines Group Representative Cinstituency, observed that the hub represented a solid public-private partnership and an integration of health and social services for the ageing population.

Tan, S. A. (2018, November 1). Gym for seniors opens in Bishan CC. *The New Paper, Singapore*, p. 4.

In collaboration with the Lien Foundation, the Bishan Community Club developed the Gym Tonic Programme, which aimed to help seniors enhance their strength and mobility as they age. Mr. Chong Kee Hiong, Member of Parliament for Bishan-Toa Payoh Group Representative Constituency, remarked that with strength and mobility enhancement, older people would gain the ability to use other exercise facilities located in the neighbourhoods.

Ang, B. (2018, November 26). New centre for seniors, caregivers. *The New Paper, Singapore*, p. 4.

A new wellness centre, GoodLife!@Yishun, was launched to offer older people age-friendly programmes and services as well as provide support to their caregivers. The centre was a joint effort between the Nee Soon South community and the voluntary welfare organisation Montfort Care.

Teng, A. (2018, December 2). More support for caregivers to seniors. *The Straits Times, Singapore*, p. A13.

Ms. Low Yen Ling, Mayor for the Southwest District, announced three new initiatives spearheaded by the Southwest Community Development Council. The initiatives were aimed at providing more financial and wellbeing support to caregivers of older adults. They included the Southwest Caregiver Support Fund, a reference guide in collaboration with the Agency for Integrated Care, and a workplace advisory developed with the Tripartite Alliance for Fair and Progressive Employment Practices and Workforce Singapore.

Rashith, R. (2018, December 20). Postmen to check in on vulnerable seniors under nationwide home visit initiative. *The Straits Times Online, Singapore*.

Singapore Post (SingPost) launched its Postman Home Visits Programme after a one-year trial period, in which postmen were trained to visit identified vulnerable older people at their homes and check on their wellbeing while fulfilling mail deliveries in the neighbourhoods. Initial service coverage included Ang Mo Kio, Henderson, Yishun, and Jurong. Through this programme, postmen managed to make friends and connect well with older residents.

Lim, A. (2019, February 14). **Seniors to get more help to stay healthy.** *The Straits Times, Singapore*, **p. A7**.

Dr. Amy Khor, Senior Minister of State for Health announced that the Ministry of Health was striving to better serve the growing ageing population in 3 key areas. These included enabling the seniors to age well, supporting more senior care services and programmes, and developing new ageing in place concepts to meet various senior needs.

Khalik, S. (2019, May 1). **Bigger polyclinic, new nursing home at Jurong health hub.** *The Straits Times Online, Singapore*.

Dr. Lam Pin Min, Senior Minister of State for Health announced that the Ministry of Health had planned to set up a new integrated health facility in Jurong for the large number of seniors residing in the region. The facility would comprise of a much larger polyclinic (compared to the current clinic) and a 700-bed nursing home. To cope with the growing ageing population and prevailing chronic illnesses, the initiative was part of the Ministry's efforts in opening more polyclinics in neighbourhoods and promoting its Community Health Assist Scheme to more beneficiaries.

2.2.4.3 Housing and Neighbourhood Development

(1982, December 16). **Home sweet homes for aged.** *The Straits Times, Singapore*, **p. 21**.

Ms. Boey Yut Mei, an executive architect with the Housing and Development Board, commented that special considerations should be paid when planning and developing public housing areas so as to accommodate the ageing population. Generally, older people expected to have easy access to community libraries, neighbourhood health centres, recreational facilities and transport systems, etc.

(1983, September 9). **'Plan housing estates with aged in mind'.** *Singapore Monitor, Singapore*, **p. 3**.

Dr. Lim Chan Yong, President of the Singapore Action Group for the Elders put forward some suggestions for the planning and development of housing areas in Singapore. In light of the ageing population, considerations should be given to community services and facilities such as adding more Senior Citizens' Clubs to residential estates, allocating ground floor units to older people with mobility issues, providing lift landing on every floor of the apartment blocks, and equipping the bathrooms with handrails.

(1984, August 4). **Downstairs toilets soon for HDB maisonettes.** *The Straits Times, Singapore*, **p. 18**.

48 maisonette flats in Bishan were the first to be upgraded by the Housing and Development Board to make them elder-friendly through improved design including

adding a toilet to the lower floor of the split-level unit. This could especially help those with mobility issues—they would not need to climb upstairs to use the toilet.

(1985, December 28). Social ties alive in HDB estates. *Business Times, Singapore*, **p. 2**.

A survey of 76 families staying at public housing flats in Ang Mo Kio indicated that even though the residents were not living in a kampong or village, they had established a lively social network among the neighbours and investing the neighbourhood with a strong sense of community. The findings also revealed that older residents, as part of the major users of neighbourhood facilities, had enjoyed social interaction with the community.

(1986, March 27). Chit-chat centres. *The Straits Times, Singapore*, **p. 10**.

Dr. Aline Wong, Member of Parliament, commented during a parliamentary debate on the Community Development Ministry that in a large number of public housing estates, senior residents had come together in groups to catch up and exchange views with each other during their leisure time. It was suggested to provide certain meeting venues and some fundamental facilities to support these senior activities.

(1986, August 26). HDB studies studio flats for elderly. *Business Times, Singapore*, **p. 2**.

The Housing and Development Board planned to launch studio apartments together with four- or five-room flats. The co-location of these flats was aimed at enabling older people to stay close to their children whilst remaining independent in their own dwelling units.

(1987, February 24). A park for all reasons. *The Straits Times, Singapore*, **p. 15**.

A 3.4-hectare park was located at a Yishun neighbourhood. It provided various venues where people of all ages could get together for exercise and activities such as tai chi lessons and cooking competitions. Older residents were reported to enjoy a lot, e.g. taking a stroll at dawn in the park.

(1987, December 8). A 'family corner' that is the talk of Yio Chu Kang. *The Straits Times, Singapore*, **p. 20**.

A void deck corner at the Yio Chu Kang public housing neighbourhood had become a second home to its nearby residents with its age-friendly facilities and pleasant environment. The residents corner was well received by the residents, especially its senior citizens.

(1988, May 7). Lifts to stop on every floor in all new HDB blocks. *The Straits Times, Singapore*, **p. 48**.

The Housing and Development Board planned to equip all future public housing estates with lift landing on every floor, allowing the old and disabled residents to have easy access and mobility within the neighbourhood.

(1988, November 21). 'Housing estates should set aside sections for the aged'. *New Paper, Singapore*, **p. 4**.

Professor Ann Wee, a member of the Advisory Council for the Aged, commented that older people should be involved in the community instead of being isolated. The provision of studio flats for older people in public housing estates as well as age-friendly services and facilities in the neighbourhood were recommended.

(1991, February 24). Void-deck toilet for Simei's elderly. *The Straits Times, Singapore*, **p. 20**.

The Bedok Town Council had built toilets at the void decks of housing blocks in Simei neighbourhood to cater to older residents who usually gathered there for coffee breaks, chit-chat and TV shows.

(1991, May 12). Amenities for aged will be within easy reach. *The Straits Times, Singapore*, **p. 17**.

The Housing and Development Board envisioned that the ageing population would be living within the communities to stay connected and share their experience with younger people. The Board planned to launch more healthcare centres and daycare centres and shorten the walking distance to amenities within the neighbourhood. The aim was to provide all required services to senior residents within easy reach.

(1992, January 26). Simple changes in flats can make life easy for the elderly. *The Straits Times, Singapore*, **p. 24**.

The Redhill Town Council general manager, Mr. Zulkifi Baharudin, commented that there was an awareness to develop or refurbish age-friendly residential properties in which older people were able to enjoy their later life while remaining in a familiar environment. The age-friendly features in dwelling unit included handles on the doors, widened bathroom entry, lower bathtub, etc.

(1992, May 9). Park for the elderly to open by year's end. *The Straits Times, Singapore*, **p. 25**.

The Tanjong Pagar Town Council proposed to build 'granny playgrounds' for older adults, with facilities such as sheltered meeting area, physiotherapy equipment and venues for gateball and petanque. The Council's plan also included encouraging businesses to contribute more to the community, e.g. offering free health screenings.

(1993, January 4). Tampines Town Council to spend $20 m on facilities. *The Straits Times, Singapore*, **p. 25**.

The Tampines Town Council planned to renovate existing neighbourhood facilities and also set up new facilities including new parks and playgrounds, meeting corners for senior people, sheltered walkways, etc. Apart from the government's efforts, the residents in the neighbourhood were encouraged to contribute more to the community in programmes such as residents' committees and neighbourhood watch scheme.

(1993, May 26). More HDB flats for elderly. *The New Paper, Singapore*, **p. 12**.

The Asian Women's Welfare Association developed the first two floors of a public housing rental block in Ang Mo Kio into new units of the Ang Mo Kio Community Home for Senior Citizens. The 24 new flats accommodated older people in need and gave them a sense of belonging to the community. In a similar vein, the Henderson Senior Citizens' Home at Bukit Merah, was located in a public housing estate to cater to senior residents and engage them in the community.

Dhaliwal, R. (1993, June 29). All HDB estates will have centres for elderly. *The Straits Times, Singapore*, **p. 17**.

The Minister for Health and Community Development announced that the government planned to set up an activity centre for senior citizens in each of the housing estate zones across Singapore. These centres would jointly serve as a supporting network for older people and provide various recreational activities and fitness sessions so as to give the older population a strong sense of belonging.

(1994, January 20). Plan for estate for elderly. *The Straits Times, Singapore*, **p. L6**.

The Salvation Army planned to launch an individual community for senior citizens that would cater to old couples who expected to live on their own while staying together with other older people. Nursing and therapy services would be provided within the community.

(1994, March 17). 'Build retirement villages in areas where elderly can be close to families'. *The Straits Times, Singapore*, **p. 22**.

The NTUC secretary-general, Mr. Lim Boon Heng advocated that affordable retirement villages for the ageing population should be located near to the residential neighbourhoods so that the retirees were able to stay connected with the rest of the community as well as live close to their families.

(1994, December 21). 'Elderly' flats upgrading. *The New Paper, Singapore*, **p. 4**.

The Housing and Development Board converted another 6 blocks of one-room rental flats in Beach Road, Toa Payoh and Redhill Close to age-friendly housing in an initiative that aimed at improving the quality of senior citizens' lives.

(1996, September 4). Villages for elderly in HDB estates will help family ties. *The Straits Times, Singapore*, **p. 23**.

The NTUC secretary-general, Mr. Lim Boon Heng again suggested to integrate retirement villages in public housing estates so that older people were able to readily access available amenities when they were in need and stay close to their families so family ties could be maintained.

(1996, December 23). Upgrading for elderly in rental flats. *The Straits Times, Singapore*, **p. 23**.

The Home Affairs Minister and the People's Action Party anchor candidate for the Bishan-Toa Payoh Group Representation Constituency, Mr. Wong Kan Seng, announced that the government would renovate 611 one-room rental flats in the neighbourhood to age-friendly units. The upgrades included building higher squatting toilets and lift landing on every floor.

Lee, J. (1997, September 27). Two in five worry about housing in old age. *The Straits Times, Singapore*, **p. 2**.

Findings from the survey of Singaporeans' attitude to public housing subsidies (commissioned by *The Straits Times* and carried out by the Singapore Press Holdings Research and Information Department) found that one in three Singaporeans surveyed preferred to live away from their children when old. In addition, 17% of those residing in public housing indicated that they wanted to move to private housing when old.

Teo, G. (1998, June 13). Flats fit for the elderly. *The Straits Times, Singapore*, **p. 60**.

The Ministry of Community Development and the Housing and Development Board would provide better living conditions to the ageing population by offering more care and support to older people in collaboration with voluntary organisations. They would enhance senior citizens' units with age-friendly features.

(1999, May 18). Accommodating the elderly. *The Straits Times, Singapore*, **p. 31**.

Elder-friendly features should be considered when designing and developing residential and public areas for the whole community so that older people would be able to grow old in the environments they felt familiar and comfortable with. The author advocated that the Housing and Development Board should consult different groups that were involved in eldercare services such as physicians, therapists, community staff and architects, etc. in their housing development.

(1999, July 27). Should all elderly folk live together? *The New Paper, Singapore*, **p. 12**.

It was reported that even though the Housing and Development Board had already provided studio flats for senior citizens, some still preferred to stay in retirement villages where they were able to enjoy the company of their peer and their children were within easy reach as well. In addition, retirement villages could provide specific facilities for the ageing population that might not be available in ordinary housing areas such as laundry, meal provision and home cleaning services.

(1999, December 3). Elderly-friendly Park for Telok Blangah? *The Straits Times, Singapore*, **p. 86**.

The Health Minister, Mr. Lim Hng Kiang announced a renovation project to upgrade the facilities in Telok Blangah neighbourhood including a 1500 m² park where residents could play tai chi and walk foot-reflexology paths after the renovation. The main purpose of the project was to improve the lives of older residents in the neighbourhood.

(2002, March 25). Everything at hand at Bedok granny flats. *The Straits Times, Singapore*, **p. H5**.

The Golden Oaks studio apartments had been specially designed for senior residents at Bedok North Street. The flats were equipped with lift landing on each floor, alert systems and other safety features.

(2005, August 29). Housing help for elderly S'poreans. *The New Paper, Singapore*, **p. 15**.

The Housing and Development Board would launch more age-friendly studio flats and renovate part of the existing one-room rental flats for the ageing population.

Sim, M. (2005, September 4). Retire to these studios. *The Straits Times, Singapore*, **p. 23**.

It was reported that age-friendly studio units were well received by older residents as they were able to live independently in the community while staying close to their families and friends. However, older adults still required more facilities and services in the neighbourhood such as laundry, nursing and emergency alert system.

(2007, March 31). Activities for elderly should be the norm HDB rules. *The Straits Times, Singapore*, **p. H1**.

It was suggested that the activities conducted by eldercare professionals and volunteers to encourage and get older people out of their homes on a regular basis were beneficial to them. These activities helped older people to cope with daily pressure and improved their mental status.

Chong, A. (2008, January 26). Transport for all, access for all. *The Straits Times, Singapore*, **p. H11**.

The Ministry of Transport announced that the transport system would be upgraded to be friendly for the whole community, especially for the ageing population and the handicapped. The upgrading initiatives included more sheltered linkways, barrier-free footpaths, wheelchair-accessible buses, lifts in mass rapid transit stations, etc.

Chua, G. (2009, December 23). Sensors to help keep an eye on lone seniors. *The Straits Times, Singapore*, **p. B2.**

Sensor alert systems would be customised to meet the needs of the ageing population and installed in the flats of three public housing blocks in the Jalan Besar neighbourhood. The sensor system could send signals to remote monitoring sites within the community when an incident happened to the older resident who was home alone.

Sudderuddin, S. (2011, November 15). Countdown to barrier-free HDB estates. *The Straits Times, Singapore*, **p. B1.**

The Housing and Development Board was planning to bring more age-friendly features to public housing estates such as barrier-free accessibility, ramps and railings to make the neighbourhood environment more appealing to the ageing population. In addition, the government would enhance the connectivity between housing areas and major public hubs like the mass rapid transit stations.

Toh, Y.C. (2012, March 8). Facilities for seniors still planned for estates. *The Straits Times, Singapore*, **p. C10.**

The government would continue to deploy elder-friendly facilities in public housing estates in order to provide an environment in which older people could grow old gracefully and with dignity. The plan included setting up more senior activity centres, offering more daycare and rehabilitation services, and training more caregivers to attend to the seniors at home. It also required the participation of the rest of the community, to build an inclusive and age-friendly neighbourhood together.

(2012, June 29). HDB towns planned with the elderly in mind. *The Straits Times, Singapore*, **p. A33.**

To build age-friendly neighbourhoods, the Housing and Development Board launched various housing schemes and policies that offered housing options for older adults. These included locating senior housing near to neighbourhood centres, providing facilities to facilitate the mobility of the ageing population, and setting up venues where older people could spend their leisure time with families and friends.

Chan, R. (2012, October 1). Support new facilities for seniors: PM. *The Straits Times, Singapore*, **p. A1.**

Over 100 facilities for older people were expected to be launched within three years. The aim was to make the neighbourhoods more accessible and wheelchair-friendly, supporting older residents with affordable healthcare and enhancing security for seniors.

(2014, April 26). Woodlands "modern kampung" to foster greater community bonding. *Today, Singapore*.

The Housing and Development Board pioneered a project named Kampung Admiralty that was aimed at establishing 100 age-friendly units together with eldercare facilities and healthcare centres. The vision was to build a more connected community immersed in the kampung (village) culture.

Au-yong, R. (2016, January 21). A few steps? HDB to make homes more elderly-friendly. *The Straits Times, Singapore*, **p. A6**.

To make public housing estates more age-friendly, the Housing and Development Board would be upgrading its Enhancement for Active Seniors (EASE) programme to cover subsidies for ramps, which are multi-step. Since its introduction in 2012, the EASE scheme had subsidised about 90,000 households to add elder-friendly features in their flats such as grab bars and slip-resistant treatment for toilet tiles.

Heng, J. (2016, January 22). HDB focus on couples, needy and elderly folk. *The Straits Times, Singapore*, **p. A9**.

New public housing development plan unveiled by the Ministry of National Development focused on the demands of the increasingly ageing population. The initiatives included smaller flat options for older people, age-friendly features for easy mobility, more upgraded estates and green parks, improved air and food quality.

Yeo, S.J. (2016, March 24). No staring into the void for these folks. *The Straits Times, Singapore*.

Yishun residents had over the years transformed a void deck senior citizens' corner into a 'kampung central', creating a space with living room, kitchen and toys. Residents would gather there to eat, play games and interact with one another.

Yeo S.J., Lim, Y. (2016, March 24). Void decks remain a vital slice of HDB life. *The Straits Times, Singapore*, **p. B16**.

Void decks were still considered a significant part of public housing life. To some older people, the void deck on the ground floor of their apartment block was a place where they could meet new friends and catch up with one another within the immediate neighbourhood, in close range of their homes.

Lee, P. (2016, April 21). Elderly residents in mature estate the focus. *The Straits Times, Singapore*, **p. A6**.

It was reported that in mature neighbourhoods, amenities serving residents' daily lives were quite well developed. These included, e.g. markets, coffee shops and transportation services. However, the residents, especially older people, were looking forward to more fitness parks to exercise.

Ho, O. (2016, May 9). More family-friendly facilities at upgraded park. *The Straits Times, Singapore*, **p. B2**.

Fitness equipment catered for inter-generational use of a family was added to the Ang Mo Kio-Hougang neighbourhood park. The Ci Yuan Sports Park had been upgraded to include new facilities such as playground, sports court, fitness equipment, etc. The older people living nearby commented that the park offered them a venue for exercise.

(2016, November 28). Tampines to get new bus interchange with elderly- and disabled-friendly features. *The Straits Times Online, Singapore.*

Tampines had a new bus interchange that was furnished with barrier-free features and priority queues with seats for older people and disabled at the concourse zone.

Au-Yong, R. (2017, December 31). $1.9b spent on sprucing up ageing HDB flats. *The Straits Times, Singapore*, **p. A3**.

The government had increased expenditures on the Housing and Development Board Home Improvement Programme (HIP) and the Enhancement for Active Seniors (EASE) Scheme, reaching around S$1.93 billion and S$40 million respectively when compared to the same period last year. The HIP was launched to address the maintenance needs of ageing properties while the EASE enabled residents to upgrade their homes with age-friendly features.

Tai, J. (2018, February 11). Open up underused spaces for the elderly. *The Straits Times, Singapore*, **p. A14**.

Designed for senior living, the 'community pocket' located at Chong Boon, Ang Mo Kio would soon become a multi-purpose space for older residents to do physical exercise, befriend one another, and enjoy their hobbies. The initiative was a collective effort between the Lien Foundation and Touch Community Services.

Lai, L. (2018, February 20). Programmes that help with seniors' needs. *The Straits Times, Singapore*, **p. B7**.

The government established several schemes to address the demands of the ageing population. One of the programmes was aimed at making homes more age-friendly to the seniors, offering up to 95% of subsidies and covering such work as non-slip flooring, grab bar instalment, wheelchair-friendly ramps, etc.

Tan, S. A. (2018, March 1). Making homes safer for elderly. *The New Paper, Singapore*, **p. 18**.

The Centre for Seniors (CFS) had set up a studio apartment, which could provide insights on making homes safer for senior residents with age-friendly features and advanced technologies such as handrails, lower light switches, sleep tracker, bathroom sensors, etc. Mr. Tan Kian Chew, Chairman of CFS, pointed out that it was important to get homes ready for ageing in place.

Choo, F. (2018, April 28). Seniors who are wheelchair users can get ramps for multi-step HDB flat entrances in second half of the year. *The Straits Times Online, Singapore.*

Mr. Lawrence Wong, Minister for National Development and Second Minister for Finance, announced that under the Enhancement for Active Seniors (EASE) Programme initiated by the Housing and Development Board, older people on wheelchairs would receive subsidised ramps, which would make their access to homes easier.

Wong, W. (2018, May 12). Elderly-friendly 'vertical kampung' with housing, healthcare facilities opens in Woodlands. *Channel NewsAsia, Singapore.*

A news coverage on Singapore's pioneering elder-friendly public housing project that was designed as a modern kampong (village)-like neighbourhood for older residents. Facilities included a community garden, community plaza, medical centres, active ageing hub, among others. Organising senior engagement activities had become more convenient with these facilities.

Seow, B. Y. (2018, May 13). Modern 'Kampung' in Woodlands opens. *The Straits Times, Singapore*, p. A2.

Kampung Admiralty, the first retirement village developed by the Housing and Development Board was opened in Woodlands with the aim of integrating senior housing with age-friendly facilities and engaging older residents in social activities and bonding events. Prime Minister, Mr. Lee Hsien Loong envisioned more such 'kampungs' to be planned and developed in the future.

Tang, L. (2018, August 19). Seniors find social support, motivated to keep fit and active at Kampung Admiralty. *TODAYonline, Singapore.*

Described as a 'modern kampong' and 'model for future public housing' by Prime Minister, Mr. Lee Hsien Loong, the Kampung Admiralty in Woodlands brought together the young and old, enabling them to jointly create community life. Other than age-friendly flats, Kampung Admiralty comprised a NTUC Health Active Ageing Hub and Admiralty Medical Centre, which aimed to provide social support and eldercare services to senior residents in the neighbourhood.

(2018, August 23). How societies should rethink eldercare with multi-generational housing. *Population.sg, Singapore.*

This article presented insights obtained from speakers at the Ageing Asia Innovation Forum. It discussed the factors that created a segregated community and advocated solutions for inclusive housing schemes that would create age-friendly housing and neighbourhoods with one-stop senior services, ageing in place and active ageing initiatives.

Goh, T., Lim, A. (2018, September 10). More gyms for the elderly, disabled in mature estates: PM. *The Straits Times, Singapore*, p. A4.

Prime Minister, Mr. Lee Hsien Loong shared the Sport Singapore's plan to set up five age-friendly gyms within developed public housing areas to help older and handicapped Singaporeans to stay active mentally and physically. The first gym would be built in Ang Mo Kio Community Centre where a large number of senior residents lived.

Liu, V. (2019, March 2). Guidelines for care in assisted living facilities. *The Straits Times Online, Singapore*.

Assisted living had been studied as a new approach to senior living by the government. Dr. Belinda Wee, founder of Assisted Living Facilities Association (Alfa), initiated an Alfa Good Practice Guide with inputs from over 40 industry professionals, from healthcare, social work, architecture and technology. The guide illustrated standards for assisted living facilities in Singapore.

Choo, F. (2019, March 7). Assisted living with care services for seniors in the pipeline. *The Straits Times, Singapore*, **p. B8**.

Dr. Amy Khor, Senior Minister of State for Health announced that a joint effort between the Ministry of Health and the Ministry of National Development was moving towards a new senior housing model. The new model would allow older people to purchase a house with elder-friendly features and targeted care services. In the pipeline was more support to caregivers of older people.

2.3 Asia-Pacific (Except Singapore)

2.3.1 South-East Asia (Except Singapore)

De Guzman, A. B., Amrad, H. N., Araullo, R. C. G., & Cheung, H. B. O. (2014). A structural equation modeling of the factors affecting an age-friendly workplace. *Educational Gerontology*, *40*(6), 387–400.

Employing structural equation modelling and grounded theory epistemology, the authors presented an empirical model, which described the impact of declining memories, wisdom and age discrimination in an age-friendly workplace. Participants (n = 200), aged between 50 and 65 years, were selected from a large government bank in the Philippines. The research method consisted of both quantitative questionnaires and qualitative interviews. No significant relationship was found between declining memories and wisdom. A positive association was found between wisdom with age-discrimination and age-discrimination and age-friendly workplaces. Recommendations included exposing older adults to newer technologies and assigning them to be mentors for younger co-workers.

Elsawahli, H., Shah Ali, A., Ahmad, F., & Al-Obaidi, K. M. (2017). Evaluating potential environmental variables and active aging in older adults for age-friendly neighborhoods in Malaysia. *Journal of Housing for the Elderly, 31*(1), 74–92.

This study utilised the WHO Age-Friendly Cities and Communities framework to investigate the relationship between built environment characteristics and active ageing for older adults. The sample consisted of 385 older adults aged 60 years and older in Malaysia. Multiple regression analysis showed that neighbourhood characteristics such as site permeability, street connectivity and accessibility to amenities resulted in higher levels of social interaction and physical activity. Findings suggested that an interdisciplinary approach that addressed multiple domains should be undertaken to improve the physical and social environment of neighbourhoods for older adults.

Fatmah, Dewi, V. P., Yudarini, & Lasmidjah, S. (2019). Readiness of Depok City to become an age-friendly city from community perspectives. *Asian Journal of Scientific Research, 12*(1), 91–96.

This cross-sectional study evaluated Depok for readiness in Age-Friendly City (AFC) in Indonesia. Some 104 pre-elderly (aged 40–59 years) and elderly (aged above 60 years) were interviewed with questionnaire on their socio-demographic, social activities and three AFC indicators including building and open spaces, housing, and civic participation and employment. Univariate analysis indicated that these three AFC indicators were considered to be lacking. The authors acknowledged that there was a great potential for Depok to become an AFC, however, policies and laws supporting AFC formation were not in place, except for an official humanitarian policy.

Fatmah, F., Dewi, V. P., & Priotomo, Y. (2019). Developing age-friendly city readiness: A case study from Depok City, Indonesia. *SAGE Open Medicine, 7,* 1–10.

This study aimed to examine whether Depok Age-Friendly City was ready in preparedness. Based on 8 age-friendly indicators developed in previous studies, 10 in-depth interviews and 4 focus group discussions were conducted to collect qualitative data from 10 and 40 key informants respectively using purposive sampling. Content analysis was applied and revealed that Depok Age-Friendly City was still in the midst of developing and not prepared as yet because of insufficient infrastructure. The authors acknowledged the indirect effect of existing regional laws and policies on age-friendliness, especially in buildings and open green spaces. They further observed that few infrastructure improvements were in place.

Lai, M. M., Lein, S. Y., Lau, S. H., & Lai, M. L. (2016). Modelling age-friendly environment, active ageing, and social connectedness in an emerging Asian economy. *Journal of Ageing Research.*

This paper adopted a structural equation model to assess the 8 key features of WHO Age-Friendly Cities among 211 caregivers and 402 self-care adults (aged 45 to 85 years and older) in Malaysia. Findings revealed that transport and housing amenities, community support and health services as well as pedestrian infrastructure were the key factors of an age-friendly environment. Short-term recommendations included reducing structural barriers of mobility to enable social interaction and reduce social isolation.

Lit, P. K. (2007). Greying Malaysians: Strategies for promoting and supporting healthy and productive ageing. *Journal of Community Health, 13.*

This paper examined public policies that promoted healthy ageing, compression of morbidity, and productive ageing in Malaysia. The method used was literature review. Since Malaysia was becoming an ageing society, the author claimed that older adults should not be seen as a burden to society and be forced into retirement. Instead, the author proposed a gradual economic and social disengagement of an older adult, which could be supported with the right policies of healthy and productive ageing. Additionally, the work, built and legal environments should become elder-friendly.

Mohammad, N. M. N., & Abbas, M. Y. (2012). The elderly environment in Malaysia: Impact of multiple built environment characteristics. *Procedia-Social and Behavioral Sciences, 49,* **120–126.**

This paper examined the relationship between the built environment and older adults' experience with their daily routines. The sample consisted of 45 able-bodied older adults aged 60 years and older in Malaysia. Older adults were asked about their preference for built environment features and design characteristics of local neighbourhoods were measured against the quality of life outcomes for participants. Findings provided an insight into the way older people utilised physical features in the built environment such as trees and buildings as well as social features, e.g. social interaction as visual cues for wayfinding.

Tiraphat, S., Peltzer, K., Thamma-Aphiphol, K., & Suthisukon, K. (2017). The role of age-friendly environments on quality of life among Thai older adults. *International Journal of Environmental Research and Public Health, 14(3),* **282.**

This study adopted the WHO Quality of Life (WHOQOL-BREF) scale to measure the significance of age-friendly environments towards the quality of life among older adults in Thailand. A cross-sectional study of 4183 older adults aged 60 years and above was conducted across four regions in Thailand. Using multivariate logistic regression analysis, walkable neighbourhood, aesthetics, accessibility to services, criminal safety, social trust, social support and social cohesion were found to be significantly associated with quality of life. Findings revealed that many domains of age-friendliness contributed to quality of life for older adults. In consequence, it was necessary to consider improvements to both the physical and social environment to enable healthy ageing.

University of Indonesia and Center for Ageing Studies. (2014). The assessment of Depok as age-friendly city (AFC). *Journal of Biosciences and Medicines, 2(06),* **5.**

This article highlighted the potential of transforming Depok, situated in metro Jakarta region, into an age-friendly city. The study involved an empirical survey of 125 older adults aged 60 years and above on the 8 WHO age-friendly domains. The findings revealed that built environment and open space were found to be lacking and indicated as important by participants. Policy recommendations included the prioritisation of providing age-friendly infrastructure and open spaces in the built environment and more collaborations between the private and public sector to rapidly achieve the vision of age-friendly city.

2.3.2 Australia and New Zealand

Aird, R. L., & Buys, L. (2015). Active aging: Exploration into self-ratings of 'being active', out-of-home physical activity, and participation among older Australian adults living in four different settings. *Journal of Ageing Research*, 501823.

A mixed-methods approach (survey, travel diary, GPS tracking) was employed to analyse the relationship between older adults' self-rated level of physical activity with out-of-home participation and physical activity. Data was gathered from 48 individuals aged 55 years and above. While a positive relationship was found between self-rated level of physical activity and time spent outdoors, no significant relationship was found between self-rated level of physical activity and time spent walking or cycling. Findings indicated that 'active ageing' policies should explicitly define the benefits and various constituents of 'active ageing' to older adults to allow them to have a holistic understanding of the concept.

Atkins, M. T., & Tonts, M. (2015). Exploring cities through a population ageing matrix: a spatial and temporal analysis of older adult population trends in Perth, Australia. *Australian Geographer, 47*(1), 65–87.

This paper examined the combination of two commonly used measures of population ageing: structural ageing (ratio of the older cohort to rest of the population) and numeric ageing (number of older people in the population) to identify spatial patterns of ageing in Perth, Western Australia. Compared to using a single measure, the combination of two measures in the form of a population ageing matrix revealed dynamic patterns underlying the spatial distribution of ageing populations. Implications for policymakers included greater precision to pinpoint areas undergoing population change and the ability to identify locational demand for age-support services to allow for more efficient appropriation of resources.

Biggs, S., & Carr, A. (2015). Age- and child-friendly cities and the promise of intergenerational space. *Journal of Social Work Practice, 29*(1), 99–112.

This paper argued for a change in the paradigm of viewing cities as places for work and consumption towards one of creativity, sense of place and social engagement. In doing so, the idea of intergenerational spaces could be shifted from one, which defined such spaces as a homogenous identity (space for all ages) to one, which embraced the heterogeneity of different identities, life-stage aspirations and considerations. It was recommended to factor in the notion of 'play' in the design of such spaces so as to allow for effective intergenerational negotiation.

Brasher, K., & Winterton, R. (2016). Whose responsibility? Challenges to creating an age-friendly Victoria in the wider Australian policy context. In T. Moulaert & S. Garon (Eds.), *Age-friendly cities and communities in international comparison: Political lessons, scientific avenues, and democratic issues* **(Vol. 14, pp. 229–246). Switzerland: Springer**.

Drawing upon Australia's (Victoria) policies, practice and academic literature, this chapter expounded the factors that accounted for limited success of age-friendly policies in the Australian context. It argued that a lack of political vision and will could be a major impediment towards policy success. To ensure successful adoption of the WHO Age-Friendly Cities framework, adequate levels of top-down leadership must accompany bottom-up approaches. This would require a paradigm shift in which policymakers recognised older adults as assets rather than liabilities and expanded their scope of policy beyond aged care and disability planning.

Broome, K., Worrall, L., McKenna, K., & Boldy, D. (2010). Priorities for an age-friendly bus system. *Canadian Journal on Aging, 29*(03), 435–444.

This paper involved a study on the barriers and facilitators of bus use for older adults aged 60 years and above in Queensland, Australia. A multi-method methodology was undertaken, involving nominal group technique (N = 227) and focused ethnography (N = 40). Based on the findings, key domains of age-friendly bus systems included the placement of vehicle entry points, bus driver friendliness, scheduling of buses, locations of bus stops, adequate pedestrian infrastructure, promoting awareness for older adults and detailing of bus routes. Recommendations to improve the bus system included focusing on poorly performing domains to overcome barriers and improving access to buses for older adults.

Han, J. H., & Kim, J.-H. (2016). Variations in ageing in home and ageing in neighbourhood. *Australian Geographer, 48*(2), 255–272.

Using research data obtained from the Household, Income and Labour Dynamics in Australia (HILDA) survey, this paper analysed the mobility decisions of older Australians aged 55 years and older and detailed the factors that they considered for ageing in place. Findings revealed that the majority (more than 96%) of older Australians preferred to age at home. In addition, older adults who were more likely to move houses also preferred to move to a new location within their current neighbourhood. The findings suggested that policies should focus on place-based initiatives such as improving existing home conditions for older adults rather than creating large-scale community care facilities like nursing homes or retirement villages.

Kendig, H., Elias, A. M., Matwijiw, P., & Anstey, K. (2014). Developing age-friendly cities and communities in Australia. *Journal of Ageing and Health, 26*(8), 1390–1414.

This paper reviewed the effect of the WHO Age-Friendly Cities framework in propagating age-friendly policies and initiatives in Australia. A review of existing literature and empirical evidence showed that the progression of age-friendly initiatives varied across cities. While cities such as Melbourne and Canberra had taken the lead to draft and implement age-friendly strategies, others such as Sydney were still working towards the development of strategies. Findings revealed that political will and fiscal austerity were critical determinants for the successful conception and implementation of age-friendly initiatives.

Kendig, H., Gong, C. H., Cannon, L., & Browning, C. (2017). Preferences and predictors of aging in place: Longitudinal evidence from Melbourne, Australia. *Journal of Housing For the Elderly, 31*(3), 259–271.

Drawn upon the Melbourne Longitudinal Studies on Healthy Ageing (MELSHA) data of 1000 community-based residents aged 65 years and older, the study examined the preferences and predictors of ageing in place. Findings revealed a positive relationship between home ownership, socioeconomic status, neighbourhood satisfaction and improvements to home with ageing in place. Residents who were renters, single adults or those who suffered from depressive symptoms were more likely to leave their neighbourhood. The study recommended including measures of age-friendly qualities of the neighbourhood to provide a more comprehensive view of older adult's decision to age in place.

Morris, A. (2013). The residualisation of public housing and its impact on older tenants in inner-city Sydney, Australia. *Journal of Sociology, 51*(2), 154–169.

This paper argued against the current public housing model in Australia, proposing a need to provide greater housing accessibility to a wider group of citizens. The current residual public housing model afforded public housing to people of lower socioeconomic status. In doing so, it reinforced the stigmatisation of these public housing estates. Drawing on a series of in-depth interviews with older public housing tenants in Sydney, Australia, the author discussed the challenges that older adults faced in navigating through their day-to-day activities while living in these public housing estates.

Vine, D., Buys, L., & Aird, R. (2012). Experiences of neighbourhood walkability among older Australians living in high density inner-city areas. *Planning Theory & Practice, 13*(3), 421–444.

This paper investigated 12 older adults' experience of walkability within high-density urban suburbs. A multi-method methodology, consisting of travel diaries, global positioning system (GPS) tracking and in-depth interviews, was undertaken. The findings revealed several key barriers to an older person's mobility: cycling paths, lack of pedestrian infrastructure and inaccessibility to transportation. The availability of green spaces was found to be an enabler for walking. The study recommended consideration of mobility, access and recreational needs of older adults when planning locations of amenities and infrastructure.

Wiles, J. L., Leibing, A., Guberman, N., Reeve, J., & Allen, R. E. (2012). The meaning of "ageing in place" to older people. *Gerontologist*, *52*(3), 357–366.

This paper explored the concept of 'ageing in place' from the perspective of older adults. A sample of 121 older adults aged between 56 and 92 years participated in focus groups and interviews in Aotearoa, New Zealand. Findings revealed that older adults had a different view of 'ageing in place' as compared to the view of the professionals. Their idea of ageing in place revolved around the concept of 'home' where it was not only limited to an individual's dwelling unit but rather the wider neighbourhood environment. In addition, the findings also revealed that older adults' view of 'ageing in place' was often homogenous in their neighbourhood and they viewed it as a way for them to maintain autonomy and independence in their neighbourhood. The research suggested that policies that hoped to achieve 'ageing in place' should be context specific.

Zeitler, E., Buys, L., Aird, R., & Miller, E. (2012). Mobility and active ageing in suburban environments: Findings from in-depth interviews and person-based GPS tracking. *Current Gerontology and Geriatrics Research*, 257186.

This paper employed a mixed method methodology, consisting of GPS tracking, travel diaries, questionnaires and semi-structured interviews to investigate the effects of the suburban environment on transport choices of 13 older adults aged between 56 and 87 years in Brisbane, Australia. Usability and accessibility of public transportation were found to be associated with the older adult's transportation choices. Findings suggested that environments conducive for active transportation not only allowed older adults to be more physically active, they also provided them with more opportunities for social interaction. It recommended that policymakers and planners should consider active mobility strategies that enabled older adults to walk and cycle safely, allowing them to remain active and engaged within their communities.

2.3.3 China

Du, P., & Xie, L. (2015). The use of law to protect and promote age-friendly environment. *Journal of Social Work Practice*, *29*(1), 13–21.

This paper examined the December 2012 revised legislation on 'Protection of the Rights and Interest of the Elderly' in China. The new legislation expanded the focus of age-friendliness from considering only structured healthcare systems to including the age-friendliness of physical environments in Chinese cities. The new law was intended to bridge the quality of life gap between urban and rural areas of China through the creation of age-friendly environments at contextual scales.

Gao, J., Fu, H., Li, J., & Jia, Y. (2015). Association between social and built environments and leisure-time physical activity among Chinese older adults—A multilevel analysis. *BMC Public Health, 15*(1).

This paper examined the relationship between leisure time physical activity with social and physical environments simultaneously. The data was drawn from a sample of 2783 older adults aged 60 years and above from Shanghai, China. Using multilevel logistic regression, individual level social cohesion, social participation, individual perceived walkability and neighbourhood walkability were found to be positively related with the prevalence of leisure time activity. The findings indicated that the creation of walkable and socially cohesive neighbourhoods could play an integrated role in helping to increase leisure time physical activity among older adults.

Gao, J., Weaver, S. R., Fu, H., Jia, Y., & Li, J. (2017). Relationships between neighbourhood attributes and subjective well-being among the Chinese elderly: Data from Shanghai. *Bioscience Trends, 11*(5), 516–523.

This paper examined the relationship between social and physical environments with subjective wellbeing among 2719 older adults aged 60 years and above in Shanghai, China. Using multilevel linear regression, a positive relationship was found between perceived neighbourhood aesthetic quality and perceived social cohesion with social wellbeing levels. Findings indicated that well-designed and closely-knit neighbourhood environments could help to promote physical activity and increase participation in social activities.

Hadjri, K., Wang, J., & Gadakari, T. (2018). Designing residential buildings for older people in China to promote ageing-in-place. *Asian Journal of Quality of Life, 3*(13), 18–28.

Building upon the WHO Age-Friendly Cities framework, this paper aimed to identify physical factors that were critical to the design of age-friendly neighbourhoods. From stakeholder interviews, several key domains were identified as critical for planning and design of age-friendly neighbourhoods: (1) accessibility, (2) sensory, (3) health and safety, (4) cognitive, and (5) social. Key barriers to community design included lack of space and fiscal constraints. Findings recommended that design solutions should be human-centric design, incorporate contextual traditional designs, provide detailed building design standards, support home modifications, create a sense of place and integrate assistive technologies into the day to day lives of older adults.

Miao, J., Wu, X., & Sun, X. (2018). Neighborhood, social cohesion, and the elderly's depression in Shanghai. *Social Science & Medicine*, *229*, 134–143.

Drawing on data from the Shanghai Urban Neighbourhood Survey, this paper investigated the association between neighbourhood environment and depression among older adults in Shanghai. Findings showed that social cohesion and engagement could be effective in managing depression rates. In addition, socio-economically disadvantaged neighbourhoods were found to have stronger social cohesion, which ameliorated the effects of depression. Recommendations included collaborating with local non-governmental organisations to promote social participation and working with urban designers and planners to provide more outdoor spaces for social activity.

Wang, Y., Chen, Y. C., Shen, H. W., & Morrow-Howell, N. (2018). Neighbourhood and depressive symptoms: A comparison of rural and urban Chinese older adults. *Gerontologist, 58*(1), 68–78.

This paper employed data from 2011 and 2013 of the China Health and Retirement Longitudinal Study to investigate the neighbourhood stressors of depressive symptoms among older adults in China. Participants included 6548 older adults aged 60 years and above. Using a multilevel modelling approach, a negative association was found between the quality of neighbourhood environments and depressive symptoms. Neighbourhood physical environment stressors affected rural respondents while social environment stressors affected urban respondents. Findings displayed the significance of the relationship between neighbourhood environment and mental health of participants over time. To improve mental health for older adults, recommendations included improving infrastructure development and accessibility for rural areas while focusing on social activities and creating a sense of place in urban areas.

Wang, Y., Gonzales, E., & Morrow-Howell, N. (2017). Applying WHO's age-friendly communities framework to a national survey in China. *Journal of Gerontological Social Work, 60*(3), 215–231.

This paper argued that the WHO Age-Friendly Cities framework did not apply equilaterally to different regions with varying levels of socio-economic development. This framework was applied to data obtained from the China Health and Retirement Longitudinal Study to measure age-friendliness of different urban and rural settings. Limited efficacy was displayed in measuring domains for developing regions, with several categories being completely omitted: (1) civic participation and employment, and (2) respect and social inclusion. Findings revealed that the WHO age-friendly cities checklist and 8 domains were heavily urban and industrially oriented, highlighting the importance of considering generalisability of indicators to lesser developed regions.

Xie, L. (2018). Age-friendly communities and life satisfaction among the elderly in urban China. *Research on Aging, 40*(9), 883–905.

Drawn upon a national sample of 9965 older adults aged 60 years and older in China, the paper examined the association between perceived built environment characteristics and life satisfaction. Using a structural equation model approach, older adults' perception of housing conditions, local amenities and social inclusion were found to be significantly associated with life satisfaction. While no significant difference was found between groups belonging to different socio-economic classes, the group that was socioeconomically disadvantaged measured lowest on perceived community age-friendliness. Findings revealed that community age-friendliness might play a significant role in affecting life satisfaction among older adults. The study recommended reducing built environment barriers or promoting enablers within the environment to increase older adults' life satisfaction.

Yan, B., Gao, X., & Breitung, W. (2016). Neighbourhood determinants for life satisfaction of older people in Beijing. In D. Wang & S. He (Eds.), *Mobility, Sociability and Well-being of Urban Living* **(pp. 231–248): Springer.**

This paper explored the importance of neighbourhood factors in influencing life satisfaction of older adults in Beijing, China. Using a structural equation modelling approach, social support was found to be the most significant factor relating to life satisfaction among urban older adults in Beijing. The study further examined the differences between neighbourhoods with a high proportion of older adults and those that had a lower proportion. The comparative analysis revealed that while community care services and amenities were a priority for the former, accessibility to key amenities was the main concern for the latter. These findings provided useful evidence to provide contextualised neighbourhood design strategies to address gaps specific to each neighbourhood.

Yan, B., Gao, X., & Lyon, M. (2014). Modelling satisfaction amongst the elderly in different Chinese urban neighbourhoods. *Social Science & Medicine, 118,* **127–134.**

Building upon the concept of 'ageing in place', this paper adopted the person-environment fit framework to assess the satisfaction of older adults and their living environment in Beijing, China. Using structural equation modelling, social support was found to be a critical factor affecting older adults' satisfaction with their neighbourhood environment. The model also differentiated neighbourhoods into four types: (1) neighbourhoods where environment and senior services were primary factors for satisfaction, (2) neighbourhoods where satisfaction was closely linked to individual characteristics such as health and income, (3) low-income neighbourhoods, which prioritised social support, and (4) neighbourhoods that required social support and better physical environments. Findings revealed the heterogeneity of the built environment and population demographics amongst urban neighbourhoods in China. It recommended that neighbourhood design strategies should be contextualised to ensure the efficacy of implementation and resource efficiency.

Yang, Q. (2016). Can we make the "New Urban Agenda" more age-friendly for island-based older people's housing? A case study of Xinzhou Town, China. *Architecture & Environment, 15*(2), **109–120.**

This paper examined the use of the UN-Habitat New Urban Agenda and the WHO Age-Friendly Cities framework to investigate housing concerns for older adults living in an island-based setting in Anqing, China. The paper argued that urban-rural linkages spurred by these frameworks had resulted in a wider range of challenges for older adults' living environment. These challenges included gentrification pressures, unavailability of support for home modifications and poor transport links that prevent access to housing services. Findings further indicated that not all the age-friendly domains identified in the WHO Age-Friendly Cities framework delivered positive outcomes for people living in developing areas. This study highlighted the need to consider specific urban contexts to allow the application of global frameworks to be inclusive to different communities.

Zhai, Y., Li, K., & Liu, J. (2018). A conceptual guideline to age-friendly outdoor space development in China: How do Chinese seniors use the urban comprehensive park? A focus on time, place, and activities. *Sustainability, 10*(10).

This paper employed an empirical approach (direct observation and questionnaire) to evaluate older adults' behavioural preference for open space in Xi'an, China. Findings revealed that Chinese older adults visited parks on a regular basis to meet with their friends and engage in physical activities or events. Activities in parks ranged from sedentary (e.g. sitting, recreational table games) to more vigorous activities (e.g. tai chi, dancing). Recommendations for the design of age-friendly urban parks included providing sufficient seating for social gatherings, basic park infrastructure (e.g. toilets, cafes, drinking fountains) and a variety of open spaces.

2.3.4   Hong Kong, SAR

Au, A., Ng, E., Garner, B., Lai, S., & Chan, K. (2015). Proactive aging and intergenerational mentoring program to promote the well-being of older adults: Pilot studies. *Clinical Gerontologist, 38*(3), 203–210.

This paper reviewed two pilot studies on proactive ageing and intergenerational programme in Hong Kong. The two interventions were evaluated in terms of their effect on older and young adults' wellbeing. Using survey, the first study, Proactive Aging Psycho-education Programme, interviewed 17 older adults while the second study, Intensive Intergenerational Mentoring Learning Experience, comprised 36 university students. In both programmes, the young and older adults interacted with each other. The authors found such programmes to be well accepted among Chinese Hong Kong residents as well as to be potentially contributing towards the social and psychological wellbeing of participants.

Au, A. M., Chan, S. C., Yip, H. M., Kwok, J. Y., Lai, K. Y., Leung, K. M., Lee, A. L., Lai, D. W., Tsien, T., & Lai, S. M. (2017). Age-friendliness and life satisfaction of young-old and old-old in Hong Kong. *Current Gerontology and Geriatrics Research, 2017*, 6215917.

This paper studied the association between the different aspects of age-friendliness as defined by the WHO and life satisfaction among young-old and old-old adults in

Hong Kong. Older adults aged 65 years and older completed a questionnaire that consisted of the Age-friendly City Scale, Satisfaction with Life Scale, and demographic variables. The authors performed multiple linear regression to test the association between age-friendliness and life satisfaction. Results showed that age-friendliness was significantly related to life satisfaction. For both groups of participants, transportation and social participation were related to life satisfaction. Additionally, for the young-old group, community and health services were significantly associated with life satisfaction whereas for the old-old group, it was civic participation and employment.

Chan, A. C. M., & Cao, T. (2015). Age-friendly neighbourhoods as civic participation: Implementation of an active ageing policy in Hong Kong. *Journal of Social Work Practice, 29*(1), 53–68.

Building upon the WHO Age-Friendly Neighbourhoods framework, this paper detailed Hong Kong's experience in improving civic participation among older adults. The approach involved one that transcended 'participation' to one where older adults were 'co-partners' in the creation of age-friendly neighbourhoods. Some of the strategies included encouraging life-long learning and volunteering within the community as well as the creation of multi-stakeholder partnerships between older adults, NGOs, government, research institutes and business sectors to organise programmes for the wellbeing and engagement of older adults. The approach called for an integrative approach of civic participation to enhance the quality of life for older adults.

Chan, A. W. K., Chan, H. Y. L., Chan, I. K. Y., Cheung, B. Y. L., & Lee, D. T. F. (2016). An age-friendly living environment as seen by Chinese older adults: A "photovoice" study. *International Journal of Environmental Research and Public Health, 13*(9), 913.

This paper described an empirical study of 44 older adults' experiential living in Chinese neighbourhoods through photovoice and semi-structured interview methods. They explored aspects of the housing environment that were considered important for ageing in place. The findings revealed three important facets of ageing in place: (1) housing design, (2) neighbourhood with supportive services, and (3) connection to family and the community. This approach engaged older people in the planning and design of their own neighbourhoods while providing insight from older adults' experience to guide the development of age-friendly neighbourhoods.

Chan, G. M. Y., Lou, V. W. Q., & Ko, L. S. F. (2016). Age-friendly Hong Kong. In T. Moulaert & S. Garon (Eds.), *Age-Friendly Cities and Communities in International Comparison: Political Lessons, Scientific Avenues, and Democratic Issues* **(pp. 121–151). Cham: Springer International Publishing.**

This chapter reported on an exploratory study to understand an age-friendly city concept in the context of Hong Kong, and to further develop expertise alliance. Eight focus group discussions were conducted with 96 participants (older adults, NGO professionals, service providers, experts) to understand their concerns and suggestions for each of the 8 WHO age-friendly domains. Most of the participants were satisfied with Transportation and Health Services whereas the domains of Outdoor Spaces and Building, and Housing were areas where older adults felt dissatisfied. Respect and Social Inclusion was another domain with which older adults were particularly dissatisfied. Recommendations were offered on how to make Hong Kong an age-friendly city.

Chan, H. M., & Pang, S. (2007). Long-term care: Dignity, autonomy, family integrity, and social sustainability: The Hong Kong experience. *The Journal of Medicine and Philosophy, 32*(5), 401–424.

This paper reported on the study of older adults, family members, administrators and healthcare professionals' perceptions on long-term care facilities in Hong Kong, with an emphasis on individual autonomy, quality and location, decision-making and financing of long-term care. In-depth, semi-structured interviews were conducted with 29 participants (care-receivers, family members, administrators, physicians and non-physician caregivers). Results showed that despite the high rate of institutional care, most participants considered a home setting or ageing in place most suitable for long-term care. Results also pointed to the importance of financial concerns and older adults' dignity in terms of respect and social connections.

Chang, J. S. H. (2014). Development of services and policy for the elderly in Hong Kong. *The Hong Kong Journal of Social Work, 48*(01–02), 65–84.

This paper reviewed the policies and services (and their challenges and achievements) for older adults in Hong Kong over the past 40 years. The author divided this time period into three eras: the budding period (before 1980), the blossoming golden period (between 1981 and 1996), and the consolidation period (between 1997 and 2012). Additionally, the author discussed the future development of Hong Kong's commitment to older adults, drawing on other countries' examples and providing recommendations such as a formulation of an expert team that would help the government plan for services. The areas where further development was possible were alleviation of poverty, retirement scheme, improvement of health, keeping older adults in the workforce, productive ageing, quality of life, lifelong learning, and life bonus.

Chau, P. H., Wong, M., & Woo, J. (2013). Living environment. In J. Woo (Ed.), *Aging in Hong Kong: A Comparative Perspective* **(pp. 31–67). Boston, MA: Springer US**.

This chapter examined the relationship between the living environment and health outcomes of older adults in Hong Kong. The authors first documented the spatial and temporal variations of various health outcomes such as stroke and case fatality across Hong Kong. To understand geographical or environmental influence on health outcomes, the authors further investigated neighbourhoods, climate, air pollution and open spaces. Additionally, a questionnaire was designed according to the

WHO Age-Friendly City guidelines and 528 respondents—almost half of whom were aged 65 years and older—took part in rating the age-friendliness across the 8 WHO domains. The authors concluded that there were still many unanswered questions regarding the relationship between the environment and health of older adults, and that more interdisciplinary research was needed to address this challenge.

Chui, C. H. K., Tang, J. Y. M., Kwan, C. M., Fung Chan, O., Tse, M., Chiu, R. L. H., Lou, V. W. Q., Chau, P. H., Leung, A. Y. M., & Lum, T. Y. S. (2018). Older adults' perceptions of age-friendliness in Hong Kong. *The Gerontologist, 59*(3), **549–558**.

In reference to the WHO Age-friendly Cities methodology, the study conducted 9 focus groups with 65 older adults to establish an age-friendly city baseline assessment in two districts in Hong Kong. The findings identified the failure of public transportation to cater to the needs of older adults, a lack of public spaces for recreation and socialising, declining human interactions in welfare services, physical and financial challenges related to housing, workplace discrimination against older people, and recommended the need to prioritise the social welfare of older adults in building a more inclusive and age-friendly city.

Chui, E. (2008). Ageing in place in Hong Kong—Challenges and opportunities in a capitalist Chinese city. *Ageing International, 32*(3), 167–182.

This paper analysed the challenges and opportunities of implementing ageing in place in Hong Kong. The author explored the specific context of rapid urban development, social changes and capitalist society in relation to ageing in place. The challenges to ageing in place included an ageing population and longer life expectancy, urban development and gentrification resulting in physical and social dislocation of older adults, and the decline of Chinese traditional values such as community and family support. Despite the challenges, the author acknowledged the many initiatives and opportunities that support ageing in place. These included the provision of community-based home care services, 'careful gentrification' to avoid displacement of older adults or minimise its consequences, safe living environment and universal design, development of housing for assisted living, and subsidy for residential care homes.

Kwok, J. Y., & Ku, B. H. (2016). Elderly people as "apocalyptic demography"? A study of the life stories of older people in Hong Kong born in the 1930s. *Journal of Aging Studies, 36*, 1–9.

This paper challenged the biomedical and economic understanding of ageing and examined the meaning of ageing through the life stories of 8 participants born in Hong Kong in the 1930s. The authors posited that everyone's experience of ageing was different, depending on one's life experiences and adopted the life-course perspective to study ageing. Participants, aged between 75 and 93 years, were interviewed three times. The interview transcripts were analysed to find similarities and differences in participants' life experiences. Results showed that older adults went through many hardships in their life but kept a sense of pride and achievement. The authors concluded that it was important to consider the life stories of older people that showed the interplay of context, history and personal experience in the ageing process.

Leung, M. Y., Yu, J., & Memari, A. (2016). Managing indoor facilities in public housing to improve elderly quality of life. *International Journal for Housing Science,* **40, 85–98.**

This paper examined the relationship between the public housing indoor facilities management and older adults' quality of life in Hong Kong. Facilities management in public housing included space planning, building services and supporting facilities. A survey was designed that measured facilities management via post-occupancy evaluation (user satisfaction) and quality of life of older adults. Some 60 older adults aged 60 years and older participated in the survey. t-test and Pearson correlation were applied to explore the association between the indoor facilities management and quality of life. Results showed that certain facilities management components such as distance, lighting, non-slip floors, and doors had a significant effect on older adults' quality of life. Recommendations were made to improve facilities management and further develop age-friendly public housing.

Mak, B., & Woo, J. (2013). Retirement and postretirement issues. In J. Woo (Ed.), *Aging in Hong Kong: A Comparative Perspective* **(pp. 69–91). Boston, MA: Springer US.**

This chapter examined retirement-related issues from the perspective of older adults as well as public and private sectors in Hong Kong. The authors explored issues such as the relevance of mandatory retirement age in a rapidly ageing society, retirement's impact on older adults' health and financial stability, public and private sector services, and post-retirement issues. The chapter presented an Elder-Friendly Employment Practice Project, CADENZA, which identified the qualities of an age-friendly workplace from the older participants' (aged 50 years and older) perspective. The results showed that many older adults in Hong Kong were willing to keep working, but they encountered many barriers in doing so. The authors also discussed the challenge of social exclusion and poverty among older adults in post-retirement age and how volunteerism was a way to avoid it.

Rozenblat, C., Wong, M., Chau, P. H., Cheung, F., Phillips, D. R., & Woo, J. (2015). Comparing the age-friendliness of different neighbourhoods using district surveys: An example from Hong Kong. *Public Library of Science One,* **10(7).**

This paper employed the WHO Age-Friendly Cities framework to assess the age-friendliness of two districts (Sha Tin and Tuen Mun) in Hong Kong. Interviews with 801 older adults aged 50 years and above were conducted based on the WHO 8 domains of age-friendliness. Both districts recorded significantly different scores. Although Sha Tin district recorded better services, infrastructure and socioeconomic status, it was perceived as less age-friendly by older adults. The findings highlighted the need to consider social and psychological factors of ageing beyond the physical environment.

Sun, Y., Phillips, D. R., & Wong, M. (2018). A study of housing typology and perceived age-friendliness in an established Hong Kong new town: A person-environment perspective. *Geoforum, 88,* **17–27.**

This paper reported on a mixed-method study that examined person-environment fit in Hong Kong. Based on the WHO Age-Friendly City concept, the study explored older adults' perceptions of the built environment and their spatial experiences in urban living. A place audit, which incorporated both qualitative and quantitative measures, was conducted in Sha Tin new town. In addition, three focus group discussions were conducted with older adults aged 65 years and above to discuss the strengths and weaknesses of their physical and social environments. Findings revealed that the physical environment was closely interlinked with the social environment. Specifically, access to positive amenities could help increase the frequency of usage and lead to higher levels of social participation and connection. The study also revealed that older adults in Hong Kong adopted a 'passive' role to ageing in place where they tried to adapt to their environment instead of activating change. Findings further suggested that older adults appreciated the built environment features such as green and open spaces, pedestrian walkways and mixed land uses. However, social exclusion and the lack of opportunities for civic participation were among the factors that older adults were less satisfied with. Recommendations included more place-making initiatives that could improve social resilience and encourage older adults to play more active roles in their communities.

Wang, D., Lau, K. K.-L., Yu, R., Wong, S. Y. S., Kwok, T. T. Y., & Woo, J. (2017). Neighbouring green space and mortality in community-dwelling elderly Hong Kong Chinese: A cohort study. *British Medical Journal Open, 7*(7).

This paper presented a longitudinal study on the availability of green space and mortality of 3544 older adults aged 65 years and older in Hong Kong. While moderating for physical activity, cognitive function, socioeconomic status, lifestyle and self-rated health, the findings revealed that a 10% increase in green space coverage was associated with a reduction in all-cause mortality, circulatory system-caused mortality and stroke-caused mortality for older adults. The findings suggested that the availability of green space could be an effective strategy for reducing mortality of older Chinese adults in highly urbanised environments.

Wong, M., Yu, R., & Woo, J. (2017). Effects of perceived neighbourhood environments on self-rated health among community-dwelling older Chinese. *International Journal of Environmental Research and Public Health, 14*(6), **614.**

In reference to the WHO Age-Friendly Cities framework, this study examined the relationship between perceived age-friendliness of neighbourhood environment with self-rated health of older adults. A total of 719 older adults aged 60 years and above participated in a questionnaire covering the 8 WHO domains. Multiple logistics regression revealed that higher satisfaction for outdoor spaces, transportation, housing, social participation, respect and social inclusion was positively related to the odds of older adults reporting higher self-rated health. However, findings also revealed that females, those with lower educational qualifications or residents of subsidised housing in the 70–79 age group were less likely to report high self-rated health. This suggested that social exclusion might be an underlying problem affecting self-rated health for this age group, regardless of individual or objective neighbourhood characteristics. The research highlighted the potential of using self-rated health as an outcome variable to evaluate age-friendliness at the neighbourhood scale.

Woo, J. (Ed.) (2013). *Aging in Hong Kong: A Comparative Perspective.* **Boston, MA: Springer US.**

This book detailed the age-friendly initiatives that Hong Kong undertook in accordance with the WHO Age-Friendly Cities initiative. The three main strategies included: (1) encourage a healthy lifestyle for people of all ages, (2) design and plan for physical environments to be age-friendly to promote social engagement for older adults, and (3) streamline existing healthcare services to be more age-friendly and intuitively accessible for older adults. The aim was to highlight the diverse range of issues that Hong Kong and other developed societies faced when working towards the goal of age-friendliness.

Woo, J. (2017). Designing fit for purpose health and social services for ageing populations. *International Journal of Environmental Research and Public Health,* *14*(5).

This paper presented the case study of Hong Kong to show how health and social services should be adopted for an ageing population and developed within the framework of the WHO concept of healthy ageing. The author documented the epidemiological data and identified weaknesses of the current Hong Kong health system such as fragmentation and lack of coordination. In order to follow the WHO proactive approach to promoting primary care for older adults and avoiding reliance on hospital system, the author proposed for a shift in health policies to ensure an integrated care of older adults in the community. For this to be achieved, a top-down approach with financial assistance for service providers would be needed.

Woo, J., Mak, B., & Yeung, F. (2013). Age-friendly primary health care: An assessment of current service provision for older adults in Hong Kong. *Health Services Insights, 6,* **69–77.**

This study evaluated the age-friendliness of primary care services for older people in Hong Kong. The assessment was carried out qualitatively in focus group discussions with older adults and service providers. The discussions followed the WHO guidelines for age-friendly primary care in areas such as information, education and training, community-based health care management systems, and the built environment. The findings were grouped in 13 categories: transport to service facility, signage, clinic facilities, physical environment, consultation process, clinic fees, community outreach service, referral system, dissemination of health knowledge, medication management, standard of care for the elderly, feedback system, and communication skills. The author recommended further initiatives such as transport modification, redesigning clinic waiting areas and toilet facilities, simplifying the appointment system, etc. to improve age-friendliness.

Yu, R., Wong, M., Chang, B., Lai, X., Lum, C. M., Auyeung, T. W., Lee, J., Tsoi, K., Lee, R., & Woo, J. (2016). Trends in activities of daily living disability in a large sample of community-dwelling Chinese older adults in Hong Kong: An age-period-cohort analysis. *BMJ Open, 6*(12), e013259.

This paper examined the trends in activities of daily living (ADL) disability in older adults aged 65 years and older in Hong Kong. The data was drawn from the Elderly Health Centres of the Department of Health that had been collecting longitudinal data on older adults since 1998. This study included the 2001–2012 data of 54808 older adults. Cross-classified random-effects logistic regression models were used to examine the trends by age-period-cohort. Results showed that ADL disability increased with age and increased more in the older age groups. Additionally, women were more likely to have ADL disability.

Yu, R., Wong, M., & Woo, J. (2018). Perceptions of neighborhood environment, sense of community, and self-rated health: An age-friendly city project in Hong Kong. *Journal of Urban Health*.

This paper utilised the WHO Age-Friendly Cities framework to assess the relationship between perceived neighbourhood environment, sense of community and self-rated health of older adults in Hong Kong. The sample involved 1798 older adults aged 60 years and older. Based on a questionnaire of the 8 WHO domains and using multivariate regression analysis, Transportation, and Respect and Social Inclusion were found to be associated with sense of community, which played an intermediary role in affecting self-rated health of older adults. Findings suggested that the relationship between neighbourhood environment and self-rated health of older adults was mediated by sense of community. The recommendation was for policymakers to focus on fostering social resiliency to improve the health and wellbeing of older adults.

Zhang, C. J. P., Barnett, A., Sit, C. H. P., Lai, P. C., Johnston, J. M., Lee, R. S. Y., & Cerin, E. (2018). Cross-sectional associations of objectively assessed neighbourhood attribute with depressive symptoms in older adults of an ultra-dense urban environment: The Hong Kong ALECS study. *British Medical Journal Open, 8*(3), e020480.

Drawing upon a sample of 909 older adults aged 65 years and above in Hong Kong, this study evaluated the associations between objectively assessed neighbourhood environment characteristics and depressive symptoms. Geographic Information Systems and environmental audit were used to evaluate objective neighbourhood environment while depressive symptoms were measured using the Geriatric Depression

Scale. Findings revealed a positive association between depressive symptoms with pedestrian infrastructure, connectivity and prevalence of public transport stops. Recommendations included optimising existing public transport fleets, reducing traffic and air-related pollution and installation green buffers to attenuate the health of residents living in highly urbanised locales.

2.3.5 Japan

Kondo, K. (2016). Progress in aging epidemiology in Japan: The JAGES project. *Journal of Epidemiology*, *JE20160093*.

This paper had two main sections. The first section introduced the Japan Gerontological Evaluation Study (JAGES), which was a panel study involving more than 100,000 participants aged 65 years and above from 2010 to 2013. The second section elaborated on the JAGES Health Equity Assessment and Response Tool (HEART), which was created to assist in implementing community-based solutions for preventive policy. JAGES surveyed older adults on health, psychological, function and social factors. Two main lines of findings emerged from the study: 1) uncovering existing disparity between fall rates and prevalence of limitation of instrumental activities for daily living (IADL) among different Japanese neighbourhoods, and 2) establishing causal evidence of negative association between increased social participation and functional decline. The JAGES HEART tool provided an intuitive framework for policymakers to analyse data indicators. The findings lend support to the benefits of a community approach, e.g. the formation of social networks and social participation to managing health policy.

Kose, S. (2019). Ageing in place: Japan's struggle towards its realisation. In A. P. Lane (Ed.), *Urban Environments for healthy ageing: A global perspective*: **Routledge**.

This chapter traced the development of housing for older adults in Japan post-Second World War. It began with the historical context of pre- and post-war housing policies and went on to address both challenges and successes of housing development. Some of the successes included the adoption of the Design Guidelines of Dwellings for the Ageing Society in 1995 while challenges included the modification of existing housing to adhere to new guidelines. Over the years, Japan had introduced several policies and laws that addressed the issue of housing for older adults. Examples included the 'Gold Plan', Securing Housing for Seniors Law, Basic Housing Law, The Housing Safety Net Law, etc. Japan had been developing solutions and innovations to support ageing in place, and the author concluded that universal design was only the starting point to achieve this goal.

To, K., & Chong, K. H. (2017). The traditional shopping street in Tokyo as a culturally sustainable and ageing-friendly community. *Journal of Urban Design, 22*(5), 637–657.

This paper discussed how community culture, social capital and older adults played an integral role in helping communities to sustain themselves. Drawing upon case studies of two traditional low-rise shopping streets in Tokyo, Japan, the paper sought to understand how these streets had managed to become microcosms of sustainability and provided resilience against urban pressures. Findings revealed that the close-knitted nature of such communities had resulted in high social capital among individuals that promoted mutual support and partnerships (elderly-serve-elderly business model, which ensured employment for older adults). In addition, these shopping streets also developed innovative business, welfare and safety models that created an authentic local cultural characteristic. The findings suggested that inclusive and bottom-up contextualised urban strategies could provide a sustainable way of urban development.

Vogelsang, E. M., & Raymo, J. M. (2014). Local-area age structure and population composition: Implications for elderly health in Japan. *Journal of Aging and Health, 26*(2), 155–177.

Drawing upon survey data of 2200 older adults aged 60 years and older from the National Survey of Japanese Elderly, this paper examined how age structure of municipalities in Japan were associated with disability at older ages as well as individual-level correlates of disability. Age structure of municipalities was defined in four categories: young ($\Leftarrow$10% older adults), average (>10% and $\Leftarrow$15% older adults), older (>15% and $\Leftarrow$20% older adults) and oldest (>20% older adults). Findings revealed that older individuals were fundamentally different between older and younger areas of Japan. While older adults from 'older' regions were more likely to have lower disability prevalence due to being married or being engaged in late-life employment, they were also more vulnerable to the onset of disability due to a lack of age-friendly infrastructure in their living environment. By identifying the compositional differences between municipalities, this paper informed the influences of different community characteristics on disability prevalence and provided important theoretical considerations for the development of age-friendly cities.

Yazawa, A., Inoue, Y., Fujiwara, T., Stickley, A., Shirai, K., Amemiya, A., Kondo N., Watanabe C., & Kondo, K. (2016). Association between social participation and hypertension among older people in Japan: The JAGES Study. *Hypertension Research, 39*(11), 818.

Employing data of 4582 older adults aged 65 years and above from the Japan Gerontological Evaluation Survey, this paper examined the association with social participation and hypertension. Using regression models, it was observed that social participation in horizontal organisations (those that adopted a non-hierarchical form of public engagement) was found to be significantly associated with reduced hypertension rate. The findings suggested that participation in horizontal organisation might increase time spent outdoors and time spent walking, which served as a pathway to reducing hypertension prevalence. Till date, a large proportion of studies were directed to examining the association of individual characteristics with individual health outcomes. This study provided supporting evidence for the consideration of a community level approach to target healthy ageing.

2.3.6 South Korea

Cho, M., & Kim, J. (2016). Coupling urban regeneration with age-friendliness: Neighborhood regeneration in Jangsu Village, Seoul. *Cities, 58*, **107–114**.

This paper examined the case of how an ageing neighbourhood (Jangsu) in Seoul managed to thrive amidst urban redevelopment pressures through bottom-up neighbourhood governance. The presence of strong social capital in Jangsu neighbourhood had enabled the creation of multi-stakeholder partnerships and inclusive processes, which allowed older adults to be part of the neighbourhood regeneration and governance process. By allowing older adults to become an autonomous stakeholder group, this had allowed Jangsu to identify many context-specific neighbourhood challenges and opportunities to drive its success. The paper highlighted the importance of social capital in creating age-friendly cities and how age-friendliness might serve as an urban regeneration framework.

Lee, K. H., & Kim, S. (2019). Development of age-friendly city indicators in South Korea. *Urban Design International*, **1–12**.

Building upon the WHO Age-Friendly Cities framework, this study developed and evaluated a set of age-friendly indicators to measure the age-friendliness of 16 Korean cities and provinces. A cross-sectional survey of 167 older adults aged 60 years and older revealed the inequality of age-friendliness across geographic regions. In addition, the degree of urbanisation had both negative and positive influence on age-friendliness. Regions with a higher degree of urbanisation recorded higher scores for health services while lower scores were attributed for economic level and social participation owing to decreased social interaction and reduced job opportunities for older adults.

Lee, M., & Kim, K. (2016). Older adults' perceptions of age-friendliness in Busan Metropolitan City. *Urban Policy and Research, 35*(2), **199–209**.

This study investigated the perceived age-friendliness of the built environment among 1000 older adults aged 60 years and above in Busan, South Korea. The evaluation was conducted based on the 8 WHO domains. Findings revealed a relatively low score of 48.1 upon 100 for the domains relating to physical and social environment. In particular, older adults that lived with their children gave the lowest rating of age-friendliness. Recommendations suggested promoting policies or programmes to improve intergenerational bonding and to establish multi-stakeholder partnerships to develop diverse cultural and recreational programmes for older adults.

Park, S., & Lee, S. (2017). Age-friendly environments and life satisfaction among South Korean elders: Person-environment fit perspective. *Aging Mental Health,* **21(7), 693–702.**

This study utilised indicators from the WHO Age-Friendly Cities framework to analyse the association of the built environment with socioeconomic status and wellbeing of older adults in Seoul, South Korea. Using multilevel regression, data from the survey of 1657 older adults aged 65 years and older revealed that the age-friendliness of housing and social inclusion was affected by socioeconomic status. The age-friendliness of transportation was found to be constant regardless of socioeconomic status. This study highlighted the complex relationship between socio-economic status and various domains of age-friendliness. It recommended that policies and programmes considered the contextual implications of the neighbourhood environment on older adults' wellbeing.

Park, S., & Lee, S. (2018). Heterogeneous age-friendly environments among age-cohort groups. *Sustainability,* **10(4), 1269.**

This study examined the association between the age-friendliness of the built environment (using the 8 WHO domains) and life satisfaction of older adults across three age-cohort groups: medium-aged (50–64; N = 2343), young-old (65–74; N = 936) and old-old (75+; N = 721). Regression models revealed that neighbourhood problems, access to public services and programs, and community engagement were significantly associated with life satisfaction across all age groups. The association was different for each age group, indicating the presence of heterogeneity in the needs of different age groups, which should be duly reflected in the planning and design of policies and interventions.

2.3.7 Taiwan

Chao, T. S., & Huang, H. (2016). The East Asian age-friendly cities promotion—Taiwan's experience and the need for an oriental paradigm. *Global Health Promotion,* **23(1 Suppl), 85–89.**

This paper assessed the progress of Taiwan's movement towards age-friendliness since it joined the WHO Age-Friendly Cities movement in 2010. The paper argued for a need to consider an oriental paradigm that aligned with the local context. The findings highlighted three areas of consideration. First, there was a need to consider collective values in the Asian context where communities valued social capital above individual interests. Second, it was important to consider the opinions of community leaders as they often played a disproportionate role in the decision-making and functioning of local communities. Third, a top-down approach to institutionalise age-friendly initiatives was preferred to rally support among different stakeholders. The paper highlighted the importance of considering different cultural contexts when undertaking social projects.

Lien, W. C., Guo, N. W., Chang, J. H., Lin, Y. C., & Kuan, T. S. (2014). Relationship of perceived environmental barriers and disability in community-dwelling elderly in Taiwan: A population-based study. *BioMed Geriatrics, 14*.

This paper presented a cross-sectional study of 200 older adults aged 65 years and above, which examined the relationship between perceived environmental barriers and disability in community-dwelling older adults. Using multinomial logistic regression, a positive association was identified between perceived environmental barriers and disability. The findings suggested the potential psychological and deterministic nature of the built environment in influencing physical activity. Recommendations included providing support for older adults to overcome perceived barriers in their built environment in order to increase their participation in physical and social activities.

Lin, W. I., Chen, M. L., & Cheng, J. C. (2014). The promotion of active ageing in Taiwan. *Ageing International, 39*(2), 81–96.

This study adopted a mixed-method approach (focus groups and mail questionnaire) to investigate how non-profit-making and community organisations promoted active ageing in Taiwan. A total of 525 (30.74%) responses were recorded out of a total of 1708 mailed questionnaires. Findings revealed that physical health, mental health, education, volunteer work, mutual respect and community participation were important factors for active ageing. Barriers for active ageing included inaccessible transportation, lack of social support, excessive distances, social isolation, and lack of safety. The findings suggested the need for an integrated approach that involved multiple stakeholders to promote active ageing. There was also a need to improve intergenerational relationships to create mutual respect and solidarity within the community.

Liu, L. C., Kuo, H. W., & Lin, C. C. (2018). Current status and policy planning for promoting age-friendly cities in Taitung County: Dialogue between older adults and service providers. *International Journal of Environmental Research and Public Health, 15*(10).

Building on the WHO Age-Friendly Cities conceptual framework, this study aimed to understand older adults' perception of current age-friendly policies in Taiwan and seek directions to inform policy improvement. This study adopted a mixed method approach consisting of a survey of 850 older adults aged 65 years and above and four focus group interviews. Findings revealed that respect and social inclusion was the highest rated domain while civic participation and employment rated lowest. In addition, findings revealed notable differences between the satisfaction levels of older adults in different regions of Taiwan, indicating a need to consider local conditions when generalising age-friendly policies to different regions at different stages of development.

Pleson, E., Nieuwendyk, L., Lee, K., Chaddah, A., Nykiforuk, C., & Schopflocher, D. (2014). Understanding older adults' usage of community green spaces in Taipei, Taiwan. *International Journal of Environmental Research and Public Health, 11*(2), 1444–1464.

This study was an exploratory analysis of seven community green space in Taipei and how older adults used these spaces. Green spaces in Taiwan provided older adults with a source of physical activity, function as a place for social interaction as well as a safe space for social and recreational activities. A mixed method approach was undertaken involving direct observation via Systems for Observing Play and Recreation in Communities (SOPARC) and 19 structured interviews with older adults. Findings suggested that community green spaces were well received by older adults and served as powerful drivers of physical and social activity participation. Recommendations included the provision of more structured and spontaneous activities at green spaces, and more age-friendly equipment at these spaces.

Shiau, T. A., & Huang, W. K. (2014). User perspective of age-friendly transportation: A case study of Taipei City. *Transport Policy, 36*, 184–191.

This study evaluated the age-friendliness of transportation in Taipei, Taiwan. The sample consisted of 610 older adults recruited at community centres for face-to-face interview. The questions included the domains of transport social welfare, convenience of transport services, quality of transit services, transport infrastructure, walking conditions and driving conditions. Using Rough Sets Theory and Importance-Performance Analysis, the data was evaluated to prioritise rankings for strategy scenarios. Findings identified targeting driver behaviour as one of the key strategies to improve the age-friendliness of existing transport system. The findings further highlighted the importance of considering not only the physical transport infrastructure but also social infrastructure.

Sun, Y., Chao, T. Y., Woo, J., & Au, D. W. (2017). An institutional perspective of "Glocalization" in two Asian tigers: The "Structure−Agent−Strategy" of building an age-friendly city. *Habitat International, 59*, 101–109.

This paper evaluated the propagation of the WHO Age-Friendly Cities initiative in a comparative analysis of Chiayi, Taiwan and Hong Kong based on the three dimensions of governance structure, agents and strategies adopted. Findings indicated that local policy framework was important in preventing knowledge silos that resulted in fragmented governance, particularly, when dealing with complex urban environment that required integrated policies. In addition, the comparison identified different modes of promoting age-friendliness in both cities. Chiayi city had a hierarchical structure consisting of much horizontal exchange of information between government and stakeholders. In Hong Kong, most age-friendly initiatives were propagated by NGOs, charities and universities. The study suggested moving towards a collaborative form of governance to develop consensus-oriented decisions.

Tsai, S. Y., Chen, T. Y., & Ning, C. J. (2016). Elderly people's social support and walking space by space-time path. *International Review for Spatial Planning and Sustainable Development*, *4*(3), 4–13.

This paper utilised the WHO Age-Friendly Cities Index to analyse the walking path and spatial activity of 22 older people in Xinyi district, Taipei, Taiwan. Using a mixed-method methodology of global positioning systems (GPS) tracking, direct observation and interviews, findings revealed three distinct groups of older adults: social, selection and necessary. The social group were older adults who use public transport and had the longest walking distance. Recommendations for this group involved creating more nodes to facilitate active mobility. The selection group were older adults that utilised public transport while having the shortest walking time. Policymakers should consider making the neighbourhood more walkable by providing trees and shelter to encourage more walking. The necessary group were older adults that spent the most time walking within the shortest distance. It was suggested that an improvement to social support could encourage these older adults to walk further. The findings highlighted the heterogeneity of older adults and raised the importance of considering the diversity of older adults in proposing design interventions.

Yeh, C. Y., Chang, C. K., & Yang, F. A. (2018). Applying a treatment effects model to investigate public amenity effect on physical activity of the elderly. *Journal of Aging and Social Policy*, *30*(1), 72–86.

This study aimed to establish the relationship between environmental attributes and physical activity among older adults through a treatment effects model. The sample consisted of 274 respondents aged 65 years and older from Taichung, Taiwan. Findings revealed that parks were positively associated with physical activity among older adults. Specifically, older adults who exercised in parks were found to have significantly higher physical activity levels than those who exercised at other locations. The findings suggested that parks might be a cost-effective way of promoting older adults' physical health in urbanised areas.

2.4   North America

2.4.1   United States of America

Barber, C. E. (2013). Perceptions of aging-friendly community characteristics: Does county rurality make a difference? *Online Journal of Rural Research & Policy*, *8*(2), 1–10.

The study examined the middle-aged, long-term residents' perception on the ageing-friendliness of their communities and if these perceptions varied according to county rurality. Fifteen focus group discussions were conducted with focus on the community factors that affected ageing and the characteristics that were critical in an aging-friendly community environment. An aging-friendly community survey was used to further assess the resident perception and their needs. Results suggested that middle-aged, long-term residents living in metro and rural-metro counties had a higher prevalence of identifying ageing-friendly community characteristic than those in rural counties, with strong emphasis on the aspects of transport provision, health care services and community connectedness.

Boufford, J. I. (2017). Advancing an Age-Friendly NYC. *Journal of Urban Health, 94*(3), 317–318.

In response to the converging trend of urbanisation and ageing population, the WHO Global Age-friendly cities project was officially launched in 2006 to identify the emerging needs of older adults and transform New York City (NYC) to be more ageing friendly. With reference to the 8 WHO domains of an age-friendly city, a comprehensive assessment was undertaken to evaluate the city's age friendliness. The assessment was conducted in different phrases and included a guided conversation with more than 1500 older adults, roundtable discussion with healthcare professionals, literature review and extensive mapping. Based on these results, 59 initiatives were commissioned to improve older New Yorkers' quality of life. The initiatives included the addition of thousands of new benches, redesign of more than 3000 bus shelters, the availability of new recreational and cultural programming for older adults. Improvements such as a 16% reduction in senior pedestrian fatalities and increased walkability among older adults were achieved.

Choi, M. S., Dabelko-Schoeny, H., & White, K. (2019). Access to employment, volunteer activities, and community events and perceptions of age-friendliness: The role of social connectedness. *Journal of Applied Gerontology*. https://doi.org/ 10.1177/0733464819847588.

This study examined the relationship between access to employment, volunteer opportunities, and community events, and older adults' perception of age-friendliness and feelings of social connectedness in Midwestern United States. Direct and indirect effects were examined. A sample of 264 older adults aged 50 years and older who participated in a survey on the 8 WHO domains was analysed. Path analysis was performed, and two main findings emerged. First, access to community events, job resources and feelings of connectedness positively predicted older adults' perceptions of age-friendliness. Second, feelings of connectedness seemed to mediate the relationship between older adults' access to community events and perception of the community's age-friendliness.

Emlet, C. A., & Moceri, J. T. (2012). The importance of social connectedness in building age-friendly communities. *Journal of Ageing Research*, 173247–173247.

Building on the WHO Age-Friendly Communities' framework, this paper detailed the empirical observation of 23 older adults (young older adults aged between 40 and 65 years, and older adults aged 65 years and older) in a community forum in Western Washington. Using the 'World Café' methodology, views were elicited on social connectivity and its relation to the creation of age-friendly communities. Three major themes emerged from the qualitative analysis of the forum findings: social reciprocity, meaningful interactions, and structural needs/barriers. Study findings reinforced the importance of social connectedness, participation and integration in maintaining age-friendly communities. The paper also recommended that the gap between research and policy implementation be bridged through multi-stakeholder partnerships.

Enguidanos, S., Pynoos, J., Denton, A., Alexman, S., & Diepenbrock, L. (2010). Comparison of barriers and facilitators in developing NORC programs: A tale of two communities. *Journal of Housing For the Elderly, 24*(3–4), 291–303.

This case study applied evaluation on a Supportive Service Program (SSP)—Living Independently in a Friendly Environment (LIFE)—in two different types of naturally occurring retirement communities (NORCs), Park La Brea and West Hollywood, Los Angeles. Qualitative analysis was employed on data from direct observation, in-depth interviews, and primary and secondary data from LIFE staff to compare similarities and differences of these two communities from five perspectives including membership, services, activities, volunteerism, and sustainability. In addition to barriers, the findings also indicated the success of LIFE in both vertical Park La Brea and horizontal West Hollywood NORCs, specifically in the area of social network and sense of community, and support for aging in place. The authors acknowledged that it was perhaps the specific type of communities that had contributed to the enablers and barriers of the SSP implementation.

Friedman, D., Parikh, N. S., Giunta, N., Fahs, M. C., & Gallo, W. T. (2012). The influence of neighborhood factors on the quality of life of older adults attending New York City senior centers: Results from the Health Indicators Project. *Quality of Life Research, 21*(1), 123–131.

This paper assessed whether neighbourhood-level factors (safety, social cohesion, and walkability) determined quality of life (QOL) of older adults in New York City. Applying multivariate binomial logistic regression analysis on data from a cross-sectional survey, 2008 Health Indicators Project (HIP) with 1660 older adults aged 60 years and above, QOL of older adults had significant relationship with safety and social cohesion, but not walkability. The authors suggested future studies on potential pathway of health, neighbourhood-influenced QOL, and also recommended urban initiatives on age-friendly neighbourhood features to improve overall wellbeing of older adults.

Goldman, L., Owusu, S., Smith, C., Martens, D., & Lynch, M. (2016). Age-Friendly New York City: A case study. In T. Moulaert & S. Garon (Eds.), *Age-friendly cities and communities in international comparison: Political lessons, scientific avenues, and democratic issues* **(Vol. 14, pp. 171–190). Switzerland: Springer.**

Building upon the WHO Age-Friendly Cities framework, this chapter examined the Age-Friendly NYC vision, which was an unprecedented public-private partnership between New York Academy of Medicine, Office of the Mayor, and the New York City Council. The vision encompassed 59 key initiatives that covered four main strategies: (1) increase utilisation of existing assets, (2) develop contextualised solutions to local problems, (3) promote financial resilience, and (4) leverage private sector knowledge and manpower. Recommendations included the development of new methods to demonstrate tangible and intangible gains to show improvement in quality of life, illustration of the increase in social capital as well as improvement in health outcomes for older adults.

Greenfield, E. A., & Mauldin, R. L. (2017). Participation in community activities through Naturally Occurring Retirement Community (NORC) supportive service programs. *Ageing and Society, 37*(10), 1987–2011.

This study aimed to understand the potential impact of formal organisations on social integration among older adults in New York City. In-depth interviews were conducted with 41 older adults to explore the determinants for their participation in Naturally Occurring Retirement Community Supportive Service Programs (NORC) activities, and their perception on social contact from these activity participations. Drawing on ecological systems theory, multi-phased analysis revealed three factors that had an impact on community activity participation (individual circumstances, programmatic contexts, and community context), and two social effects (limited changes, profound changes) from these participations. The authors acknowledged the influence of formal organisations on social integration in later life through multiple determinants.

Guo, K. L., & Castillo, R. J. (2012). The U.S. long term care system: Development and expansion of Naturally Occurring Retirement Communities as an innovative model for aging in place. *Ageing International, 37*(2), 210–227.

This paper elaborated the implementation of naturally occurring retirement communities (NORCs) in terms of benefits, facilitators and challenges in the United States. Results also provided evidence that NORCs, an innovative and viable approach for aging in place, functioned as both a formal and informal home for older adults to improve their health and mental wellbeing, and could be improved with provision of services and support efficiently and cost-effectively. The authors acknowledged the policy implication towards more liveable communities, and recommended future case studies with older adults in NORCs, which might address the needs of older adults and contribute to healthy and successful aging.

Hand, C. L., & Howrey, B. T. (2017). Associations among neighborhood characteristics, mobility limitation, and social participation in late life. *The Journals of Gerontology: Series B, 74*(3), 546–555.

This study examined the association of neighbourhood characteristics and its interaction with mobility limitation on social participation of three activities among older adults in the United States. Multivariate logistic regression models were deployed on data from community dwelling older adults aged 65 years and older (N = 3985–3995) in 2008 Health and Retirement Study, and census data obtained from the American Community Survey. Results supported the hypothesis that more social participation was associated with higher percentage of older residents in all three activities, greater social cohesion in nonreligious meeting, and higher population density in club. In addition, interaction effects of walking limitation with population density or social cohesion were found to be associated with more participation in some activities. The authors suggested that the interaction between neighbourhood and individual characteristics should be considered in policy, practice and future research on the development of age-friendly neighbourhoods.

Hong, A., Sallis, J. F., King, A. C., Conway, T. L., Saelens, B., Cain, K. L., Fox, E. H., & Frank, L. D. (2018). Linking green space to neighborhood social capital in older adults: The role of perceived safety. *Social Science & Medicine, 207,* **38–45**.

This paper examined whether the perception of traffic, pedestrian and personal safety was associated with green space and social capital among community-dwelling older adults in Seattle-King County and Baltimore-Washington DC region. Some 647 older adults aged 65 years and above were recruited from the Senior Neighborhood Quality of Life Study. Survey was administered to measure the social capital, green space exposure and perceived safety of the neighbourhood. Multilevel model and interaction plot were fitted. Green space features were found to impact positively on the social capital among older adults in the neighbourhood. Certain types of green spaces including parks and trees were perceived as unbeneficial for older adults who perceived their neighbourhood as unsafe. Findings suggested that pedestrian safety played a significant role in the association between green space and social capital among older adults, and further research should be conducted to confirm the causality.

Hrostowski, S. (2010). Diversity in aging America: Making our communities aging friendly. *Race, Gender & Class, 17*(3/4), **307–313**.

This paper elaborated the development of Aging-Friendly Community in the United States as well as changes and enhancements identified at the Issues on Aging Conference. Drawing on 5 of the 8 WHO domains of age-friendliness—Housing, Transportation, Recreation, Civic engagement, and Access to services—both experts and lay people attending the conference identified the actions towards age-friendly communities. Results suggested that age-friendly actions included public and government education on ageing population and advocacy attempts on physical and social infrastructure enhancement. The paper highlighted the importance of community involvements in the process of developing age-friendly communities, which would advantage people from all stages of life and environments.

Hunter, R. H., Anderson, L. A., Belza, B., Bodiford, K., Hooker, S. P., Kochtitzky, C. S., Marquez, D. X., & Satariano, W. A. (2013). Environments for healthy aging: Linking prevention research and public health practice. *Preventing Chronic Disease, 10,* **E55–E55**.

This paper explored the environmental determinants of health and described the experience of translating knowledge into practice from the Healthy Aging Research Network framework initiatives in the United States. A Knowledge to Action framework was applied to inform age-friendly environmental practice and policy by emphasising sustained, multidisciplinary, cooperative strategy and continuous involvement of researchers and communities. The experience provided evidence of scientific and practical approaches of potential environmental factors contributing to health and wellbeing of older adults and communities.

John, D. H., & Gunter, K. (2016). EngAGE in community: Using mixed methods to mobilize older people to elucidate the age-friendly attributes of urban and rural places. *Journal of Applied Gerontology, 35*(10), 1095–1120.

This paper studied both urban and rural environmental features that influenced the perception of age-friendliness and aging-in-place in the United States. Explanatory sequential mixed methods were employed using 'People and Places' model modified from the WHO Age-friendly City framework. Information on perceived availability and importance of age-friendly community features and health behaviours was collected from a quantitative telephone survey with 387 adults aged between 21 and 93 years from Clackamas County. This was followed by qualitative Mapping Assets Using Participatory Photographic Surveys (MAPPSTM) with 237 respondents aged 50 years and older in six communities of Oregon county to explore the facilitators and challenges towards age-friendly communities. Integration of results from both methods revealed similarities and variation of perceived age-friendliness independent of rural and urban communities. The authors endorsed the implications of participatory, collaborative approaches on age-friendly programmes and policy.

Lehning, A. J. (2014). Local and regional governments and age-friendly communities: A case study of the San Francisco Bay Area. *Journal of Aging & Social Policy, 26*(1–2), 102–116.

This study presented a case study to evaluate the presence of age-friendly policies, programmes and infrastructures in local governments of the San Francisco Bay Area. Data was obtained from the literature review and online surveys with local government experts from 5 of the 8 WHO domains including Community design, Housing, Transportation and mobility, Health and supportive services, and Community engagement. Descriptive analysis showed that more supportive features with close substitute mobility options (e.g. privileges for mixed-use neighbourhoods, encouragement on accessible public transit), but limited policies and programmes to encourage driving for older adults were in place. Findings indicated that the age-friendly policies, programmes and infrastructure from local and regional government and efforts of other institutions could reinforce each other. Further research could focus on exploring the roles of governmental stakeholders in the development of age-friendly communities.

Lehning, A. J., Smith, R. J., & Dunkle, R. E. (2014). Age-friendly environments and self-rated health: An exploration of Detroit elders. *Research on Aging, 36*(1), 72–94.

This paper focused on assessing the relationship between age-friendly environmental features and self-rated health among older adults in the City of Detroit, Michigan. Information of age-friendly environmental characteristics, health outcomes, and demographics were obtained from the Detroit City-Wide Needs Assessment of Older Adults with representative samples of community dwelling adults aged 60 years and older. Linear regression models demonstrated more accessible health care, social support, community engagement, and less neighbourhood problems had significant relationship with better self-rated health, after taking into account the demographic and health characteristics. After adjusting age-friendly environmental factors, income and education did not have any impact on self-rated health anymore. The authors suggested the need to study the effects of age-friendly environments, including physical and social environments on self-rated health and other outcomes (e.g. quality of life, life satisfaction, ageing in place) as well as the differences of these effects in different contexts and populations in future research.

Lehning, A. J., Smith, R. J., & Dunkle, R. E. (2015). Do age-friendly characteristics influence the expectation to age in place? A comparison of low-income and higher income Detroit elders. *Journal of Applied Gerontology, 34*(2), 158–180.

This research studied the age-friendly environmental factors associated with expectation to age in place, and whether there was variation of these associations between older adults with different income levels in Detroit, Michigan. Logistic regression models and matched pair analysis were applied on a representative data collected from 1766 community dwelling adults aged 60 years and above who completed Detroit Citywide Needs Assessment of Older Adults. Results suggested that neighbourhood problems were the only factor that had significant association with expectation to age in place among the six age-friendly characteristics identified by the US Environmental Protection Agency's age-friendly guide. Older adults with low-income level tended to have higher expectations to age in place as compared to those with higher income. The authors highlighted the possibilities of being unable to leave for contributing to ageing in place and suggested to concentrate on the role of financial resources on options and expectations of ageing in place in future studies.

Loukaitou-Sideris, A., Levy-Storms, L., Chen, L., & Brozen, M. (2016). Parks for an aging population: Needs and preferences of low-income seniors in Los Angeles. *Journal of the American Planning Association, 82*(3), 236–251.

This study aimed to gain deeper understanding of low-income older adults' needs and perceptions on parks in Los Angeles. Information on facilitators, barriers, and preference of using parks were generated from literature review and five focus groups with 39 low-income and multi-ethnic older adults aged between 62 and 91 years and residing in inner-city neighbourhood. Challenges for using parks as identified by older adults included: (1) insufficient social activity programmes in parks and along linkage paths that addressed safety and security concerns for older adults; (2) limited exercising and walking chances; (3) inadequate aesthetic and natural surroundings. Four strategies for senior-friendly parks were recommended to address these challenges. The authors also suggested future research to focus on the variation of park preferences between older adults from various ethnic backgrounds.

Maltz, J., Hunter, C., Cohen, E., & Wright, S. (2014). Designing for a lifetime in New York and other US cities. *Architectural Design, 84*(2), 36–45.

This paper described the age-friendly programmes in New York and other US cities. In addition to the groundwork of the WHO Age-friendly Cities Programme, policies and practices such as innovative centres for older citizens, physical environment enhancement, affordable housing for older adults were developed and implemented. Experiences learned from these initiatives included age-friendly features of multigenerational communities, navigation improvement, diverse services/facilities provision, and buildings for interaction between communities. Pedestrian-oriented communities in high-density cities for all stages of life were recommended for ageing in place, independently and safely.

Michael, Y. L., & Yen, I. H. (2014). Aging and place—Neighborhoods and health in a world growing older. *Journal of Aging and Health, 26*(8), 1251–1260.

This paper featured the existing efforts and research to recognise the needs for place-based strategies to improve health among older adults. A roundtable workshop 'Ageing and Place—Neighborhoods and Health in a World Growing Older' was conducted in San Diego. Attended by researchers and major stakeholders, the workshop focus was on racial minority older adults, research methodologies and policy related to age-friendly communities. Health disparities were found among the different races. There was no clear disparity in the area of social connectedness among the different races. Models included socioeconomic model, psychosocial stress model and structural-constructivist models that incorporated aspect of place in relation to health disparities. Research should focus on the neighbourhood mechanisms for health and investigate the impact of ageing- and place-related policies and intervention.

Parekh, R., Maleku, A., Fields, N., Adorno, G., Schuman, D., & Felderhoff, B. (2018). Pathways to age-friendly communities in diverse urban neighborhoods: Do social capital and social cohesion matter? *Journal of Gerontological Social Work, 61*(5), 492–512.

This paper examined the impact of social capital and social cohesion on pathways in constructing age-friendly communities among older adults. Data from the Community-Based Participatory Research Study were used. Interviews and focus group discussion were conducted with older adults aged 55 years and older to understand lived experience and perception on social identity, social connectedness and civic engagement in an age-friendly community. Majority of the older adults were able to engage in civic engagement in building age-friendly communities due to their social networks and resources. Social, cultural and linguistic barriers were observed among certain ethnic population such as the Hispanic. These findings demonstrated that social cohesion and capital had a significant role to play in constructing an age-friendly community that deepened the social connectedness and civic engagement among older adults.

Scharlach, A. (2012). Creating aging-friendly communities in the United States. *Ageing International, 37*(1), 25–38.

The author described the different types of age-friendly efforts that were initiated in the country and how the different stakeholder could play a role in these initiatives and further development. Four domains of age-friendly initiatives, derived from a national survey, were identified: community planning, system coordination, service co-location, and consumer association. The limited government rights and financial resources were some of the major barriers that most community interventions faced. Although moving towards the private sector as a solution seemed to be increasing, this was not significantly accessible. Overall, sustainability and resource availability remained as a challenge in meeting the increasing needs of an ageing community.

Scharlach, A. E. (2017). Aging in context: Individual and environmental pathways to aging-friendly communities—The 2015 Matthew A. Pollack Award Lecture. *The Gerontologist, 57*(4), 606–618.

This paper examined three important aspects of an ageing-friendly community: interaction between aging and environment; conceptualising theory with developmental processes and environmental context; and identifying implication on developing ageing-friendly communities. The author posited that older adults' wellbeing promotion was one of the primary purposes of ageing-friendly communities derived from the existing healthy ageing concepts. Environmental pathways, which fostered wellbeing among older adults along their life course were drawn from environmental gerontology principles. The author proceeded to propose an integrative Process Model of Constructive Ageing. This model put together evidences from healthy ageing interventions, phenomenological perspective, strategies and ecological frameworks. It also reflected the concept behind person-environment interaction and the adaptational process between individuals and their environment over time.

Scharlach, A. E., Davitt, J. K., Lehning, A. J., Greenfield, E. A., & Graham, C. L. (2014). Does the village model help to foster age-friendly communities? *Journal of Aging & Social Policy, 26*(1–2), 181–196.

This article examined a social initiative village model that focuses on members' involvement and accessibility to help promote ageing-friendliness among communities. This study used data from village members who participated in a national study on community-based initiatives. A survey, which incorporated the 8 WHO domains, was developed to explore the activities that could potentially help members to better access supports and services. Findings suggested that social organisation had a significant role to play in supporting older adults' ability to age in place.

Scharlach, A. E., & Lehning, A. J. (2012). Ageing-friendly communities and social inclusion in the United States of America. *Ageing and Society, 33*(01), 110–136.

This paper explored the different aspects—physical and social infrastructure that could play a role in making the communities more ageing-friendly and socially inclusive. The authors defined ageing-friendly communities as a place where physical and social environment provided age-related support to all individuals and enabled them to age in place. It should also promote social inclusiveness among individuals, which could help to overcome barriers that stopped older adults from social participation.

Smith, R. J., Lehning, A. J., & Dunkle, R. E. (2013). Conceptualizing age-friendly community characteristics in a sample of urban elders: An exploratory factor analysis. *Journal of Gerontological Social Work, 56*(2), 90–111.

This study examined the extent of social and physical environmental factors contributing to an age-friendly community. Exploratory factor analysis was conducted to reflect on the existing U.S. Environmental Protection Agency (EPA) policy framework on older adults. The EPA framework combined the principles on smart growth and active ageing. The findings indicated that accessibility to business and leisure, social interaction, accessibility to healthcare, neighbourhood problems, social support and community engagements were significant measures for age-friendly community. The results underscored the importance for future research work to revisit the influence of the environment to better understand the association between social and physical community on individual outcomes.

Smith, R. J., Lehning, A. J., & Kim, K. (2017). Aging in place in gentrifying neighborhoods: Implications for physical and mental health. *The Gerontologist, 58*(1), 26–35.

This study examined the impact of neighbourhood modification on older adults, especially those who were financially unstable. Data from the National Health and Ageing Trends Study and the 1970–2010 National Neighborhood Change Database were used. A quasi-experimental approach was used to estimate the effect of neighbourhood gentrification on self-rated health and mental health. Findings suggested that financially unstable older adults who lived in a gentrifying neighbourhood had a higher self-rated health as compared to those who lived in a low-income neighbourhood. More depression and anxiety symptoms were found among the financially unstable and higher-income older adults who lived in gentrifying neighbourhoods than those living in more affluent neighbourhoods. Poorer mental health was found among higher-income older adults who were in gentrifying neighbourhood than those in low-income neighbourhoods. These findings indicated the complexity and significance of gentrification impact and demonstrated the importance of community initiatives and future research on the influences of neighbourhood characteristics on the health of older adults.

Spring, A. (2017). Short- and long-term impacts of neighborhood built environment on self-rated health of older adults. *The Gerontologist, 58*(1), 36–46.

This study examined how self-favourable built environment and long-term exposure of deteriorated environment impacted health through a longitudinal approach. Data from the Panel Study Income Dynamics, 1999–2013, were used. Respondents were Americans aged 45 years and older who were interviewed biennially during the survey wave. Conventional logistics regression models were used to predict self-rated health at baseline with time-varying covariates. Residential self-selection bias was controlled by comparing the conventional logistic regression models to marginal structural model and inverse probability treatment weights. Findings indicated that neighbourhoods that lacked health-related facilities and were commercially declining would lead to an increased risk of poor self-rated health in the long run. Improving accessibility to health-related facilities and services was important for individuals before they reached old age.

Stokes, J. E., Morelock, J. C., & Moorman, S. M. (2016). Mechanisms linking neighborhood age composition to health. *The Gerontologist, 57*(4), 667–678.

This study examined if intergenerational factors, daily decimation and/or social well-being played a role in the association between neighbourhood age composition, health and psychological wellbeing among participants aged between 30 and 84 years. Data from the National Survey of Midlife Development in the United States 2006 Wave and the 2010 U.S. Census on neighbourhood characteristics were used. Neighbourhoods that had a similar age distribution to the United States and neighbourhoods that had a higher proportion of older adults reported more generativity and social cohesion. Residents with higher generative and social cohesion were related to better physical and psychological wellbeing, suggesting that these neighbourhood types would be able to achieve maximal benefits from ageing in place related initiatives.

Tang, F., Xu, L., Chi, I., & Dong, X. (2017). Health in the neighborhood and household contexts among older Chinese Americans. *Journal of Aging and Health, 29*(8), 1388–1409.

This study examined how neighbourhood features and living arrangement could be associated with the physical and mental wellbeing of the Chinese American older adults. Data on health, socio-demographics, living arrangement, social aspect and neighbourhood characteristics were collected from community-dwelling older adults from Greater Chicago. Results were derived through multinomial logistic, Poisson and negative binominal regression analyses. The exposure to threatening neighbourhood disorders such as safety-related issues and poor maintenance was found to be negatively associated with participants' physical and mental wellbeing. Living arrangement was found to be an important aspect as it was shown that there was more significance between social cohesion and mental wellbeing for individuals who lived with their spouse, children and/or grandchildren as compared to those who lived with spouse only. In order to improve the wellbeing of the Chinese older adults, the authors suggested for relevant policies and interventions to be in place to enhance neighbourhoods' physical and social aspects.

Yeh, J. C., Walsh, J., Spensley, C., & Wallhagen, M. (2016). Building inclusion: Toward an aging- and disability-friendly city. *American Journal of Public Health, 106*(11), 1947–1949.

The authors of this editorial explored the Aging and Disability Friendly San Francisco (ADF-SF), which aimed to influence local changes through social related public health approaches that were inclusive of older adults and disabilities personnel. ADF-SF gathered data through focus group discussions, community roundtables, stakeholder events and city-level report to assess the ageing and disability friendliness of San Francisco. A framework was conceptualised through conditions and social aspects that were affecting the wellbeing of the residents. Nine domains including outdoor spaces and building, transportation, housing, social participation, respect and social inclusion, civic participation and employment, communication and information, community and health service, and technology were identified. This framework could potentially be used to recognise community-led solutions that were related to social changes.

2.4.2 Canada

Brooks-Cleator, L. A., Giles, A. R., & Flaherty, M. (2019). Community-level factors that contribute to First Nations and Inuit older adults feeling supported to age well in a Canadian city. *Journal of Aging Studies, 48*, 50–59.

This paper examined whether the First Nation and Inuit older adults felt supported to age well in Ottawa, Canada. Using community-based participatory research to query the question—'what community-level factors contribute to indigenous older adults aged 55 and above feeling supported to age well in the city of Ottawa?'-thematic analysis was performed on the results from semi-structured interviews, focus groups and photovoice interviews with 32 indigenous older adults aged between 55 and 79 years. Social and physical environments, more specifically, health and community services by indigenous organisations and accessibility were found to support participants to age well. Areas such as housing, information, transportation, respect and recognition for indigenous older adults were identified as those that did not fully support and could be better developed to support indigenous older adults to age well.

Fang, M. L., Woolrych, R., Sixsmith, J., Canham, S., Battersby, L., & Sixsmith, A. (2016). Place-making with older persons: Establishing sense-of-place through participatory community mapping workshops. *Social Science & Medicine, 168*, 223–229.

This study assessed the factors influencing affordable housing development by participatory community mapping workshops (PCMWs) with older adults and stakeholders in Western Canada. Applying two-step methods, experiential group walks and mapping exercise sequentially, four PCMWs were conducted with 38 older participants aged 60 years and above, and 16 service providers from government agencies to co-construct and explore visual representation of older adult's perception and interaction with the place as well as to identify activity, service, physical and social environmental factors that promoted ageing in place. Results from thematic analysis showed that identifying services and needs for health and wellbeing, having opportunities for social participation and overcoming cross-cultural challenges were determined to enhance positive aging-in-place. The authors acknowledged PCMWs as a grounded approach to encourage community empowerment and development.

Finlay, J., Franke, T., McKay, H., & Sims-Gould, J. (2015). Therapeutic landscapes and wellbeing in later life: Impacts of blue and green spaces for older adults. *Health & Place, 34,* **97–106.**

Drawing on interview data with 141 older adults in Vancouver, Canada, this article examined the relationship between therapeutic landscapes with older adults' health. Interview data were categorised through a qualitative framework on physical, social, and mental health dimensions. Findings highlighted how therapeutic landscapes provoked feelings of restoration, renewal and spiritual connectedness while also serving as a multi-generational space for social engagement. Walkability was mentioned as a necessary factor for ensuring accessibility to therapeutic spaces. Policy considerations included ensuring that the cost of access to therapeutic landscapes did not serve as a barrier for older adults and small-scale green plots located in close proximity could go a long way towards improving older adults' health.

Garon, S., Paris, M., Beaulieu, M., Veil, A., & Laliberté, A. (2014). Collaborative partnership in age-friendly cities: Two case studies from Quebec, Canada. *Journal of Aging & Social Policy, 26***(1–2), 73–87.**

This paper illustrated the implementation process of age-friendly cities in Quebec (AFC-QC), Canada, and identified the facilitators and barriers for its effectiveness from a collaborative partnership perspective. A three-step implementation was adopted including social diagnostic of older adult's needs, development of logic model as an action plan, implementation through collaborations. Applying a community-building approach on two cases studies, data was collected and triangulated using mixed methods with a 5-year longitudinal design involving older adults and other stakeholders. Results demonstrated the importance of the collaborative partnership with steering committee and older adults and revealed the limited capacity of the implementation of AFC-QC independently. While acknowledging the need for an upstream top-down model, the authors suggested additional bottom-up approach with collaborative partnership in the development and evaluation for AFC initiative.

Garon, S., Veil, A., Paris, M., & Rémillard-Boilard, S. (2016). How can a research program enhance a policy? AFC-Quebec governance and evaluation opportunities. In T. Moulaert & S. Garon (Eds.), *Age-friendly cities and communities in international comparison: Political lessons, scientific avenues, and democratic issues* **(Vol. 14, pp. 99–120). Switzerland: Springer.**

Adapting the WHO Age-Friendly Cities and Communities framework as well as logic models from the Canadian International Development Agency, this chapter evaluated the AFC-Quebec model, which was based on a methodological approach of community development planning. The model emphasised the participation of older adults as a vital element of age-friendly policies and focused on creating collaborative partnerships that were managed by a steering committee. Three main steps to the Quebec model included: (1) social diagnostic involving investigation of older adults' needs and profiling of services for seniors, (2) action planning by sorting priorities and examining feasibility and relevance of initiatives and (3) implementation of proposed initiatives. The paper also discussed how the linkage of research projects with governmental programmes helped to inform policymaking and produce emergent research findings.

Garvin, T., Nykiforuk, C. I. J., & Johnson, S. (2012). Can we get old here? Seniors' perceptions of seasonal constraints of neighbourhood built environments in a northern, winter city. *Geografiska Annaler. Series B, Human Geography, 94*(4), 369–389.

This study explored the seasonal constraints of northern urban environments in Edmonton, Alberta, Canada from the perspective of older adults. Using photo-elicited method and focus group discussion, a pilot study was conducted with 11 participants aged 55 years and older. Deductive analysis, applying the WHO Age-friendly Cities framework, inductive analysis and interpretive analysis identified the environmental elements relevant to 5 age-friendly city domains (outdoor spaces and buildings, transportation, social participation, respect and social inclusion, communication and information). Additional characteristics were also identified including noticeable consideration for the winter season, different preferences of built environment in summer and winter, conflation of public and private spaces in winter. Another finding was that older participants made an effort to accommodate themselves to environmental challenges due to mobility decline. The public health implications included the need to provide safe and vibrant neighbourhood supportive environments for older adults through upstream strategies, especially in winter cities.

Hirsch, J. A., Winters, M., Clarke, P. J., Ste-Marie, N., & McKay, H. A. (2017). The influence of walkability on broader mobility for Canadian middle aged and older adults: An examination of Walk Score™ and the Mobility Over Varied Environments Scale (MOVES). *Preventive Medicine, 95*, S60–S67.

This paper described the mobility of older adults in British Columbia and assessed the relationship between built environment and mobility. Using data of 2046 participants aged 45 years and older from a population-based cross-sectional study, Canadian Community Health Survey-Healthy Aging (CCHS-HA), Mobility over Varied Environments Scale (MOVES) was developed as a holistic measurement tool of mobility that included physical, cognitive, social health and transportation. Descriptive analysis showed that the mean score for MOVES scale was 30.67 among the study population. Although preliminary analysis showed socio-demographic factors (younger, married, higher socioeconomic status, and better health status) contributed to individual mobility, results from linear regression found MOVES increased 4.84 and 0.10 points for every 10 more points in walkability scale (Street Smart Walk Score™) before and after adjusting covariates respectively. This study provided evidence on the linkage between walkability and mobility, which had implication on neighbourhood design to improve health of middle and older adults. The authors suggested future studies to further understand the association between physical and social environmental determinants and individual mobility.

Keating, N., Eales, J., & Phillips, J. E. (2013). Age-friendly rural communities: Conceptualizing 'best-fit'. *Canadian Journal on Aging, 32*(4), 319–332.

This study assessed the reconceptualised WHO definition of 'age-friendly' among older adults in rural communities to explore older person-environment fit in Canada. Two case studies were conducted in bucolic and bypassed rural living environments to test the refined age-friendly concept. Directed content analysis was performed on archival materials, census data, semi-structured interviews and group consultations to discuss the needs and resources of marginalised and community-active older adults. Findings offered evidence of the diverse locality needs and resources, and dynamic and interactive person-environment changes over time. The author also indicated the potential of reconceptualising the WHO age-friendly concept.

Krawchenko, T., Keefe, J., Manuel, P., & Rapaport, E. (2016). Coastal climate change, vulnerability and age friendly communities: Linking planning for climate change to the age friendly communities agenda. *Journal of Rural Studies, 44*, 55–62.

This study evaluated the social and place vulnerability of rural ageing communities, specifically to coastal climate change, in addition to the development of age-friendly communities in Nova Scotia, Canada. Using mixed method study design, spatial analysis was conducted on Statistics Canada Census data and GIS mapping of flooding impacts, residential housing, and public infrastructure and asset mapping at Lunenburg County and Annapolis Valley study areas. Content analysis was performed on official plans, strategies and policies relevant to physical and social environment in each study site. Results showed increasing trends of ageing population, influences of tide and storm surge on important community assets for older adults, and under adaption of municipal planning to address climate change across the study areas. Even though the findings emphasised the importance of proactive planning practices, considering the environmental hazards on age-friendly community initiatives, the author pointed out that there was a lack of concentration on place vulnerability.

Levasseur, M., Desrosiers, J., & St-Cyr Tribble, D. (2008). Do quality of life, participation and environment of older adults differ according to level of activity? *Health and Quality of Life Outcomes, 6*(1), 30.

This study examined whether quality of life, participation and perceived quality of environment differed among community-dwelling older adults with different level of activity in Canada. Descriptive analysis, chi-square test, analysis of variance (ANOVA), Welch F-ratio, and regression analysis were performed on data from a cross-sectional study, which collected information of quality of life (Quality of Life Index), participation (Assessment of Life Habits), and environment (Measure of the Quality of the Environment) with 156 older adults aged 60 years and above.

Results showed that quality of life and satisfaction with participation were positively associated with higher level of activity. After controlling for covariates, these differences were only observed between those without activity limitations and those with moderate to severe activity limitations. The study also showed that more limitation on activity level led to more restricted participation, and in turn more perceived obstacles in the physical environment. Older adults with different activity level did not demonstrate difference in facilitators in physical and social environment or obstacles in social environment. The authors recommended to assist older adults' ageing in place with more than activity.

Levasseur, M., Dubois, M. F., Généreux, M., Menec, V., Raina, P., Roy, M., Gabaude, C., Couturier, Y., & St-Pierre, C. (2017). Capturing how age-friendly communities foster positive health, social participation and health equity: A study protocol of key components and processes that promote population health in aging Canadians. *BMC Public Health, 17*(1), 502.

This paper presented a mixed-method sequential explanatory design to identify which and how the key determinants of age-friendly communities promoted health, social participation, and health equity across communities in Canada. Quantitative data were collected from 3555 participants in a survey of Canadian communities on 7 age-friendly key domains (physical environment, housing options, social environment, opportunities for participation, community supports and healthcare services, transportation options, communication and information) while information of health, social participation, health equity were obtained from 41,085 participants aged between 45 and 85 years in the Canadian Longitudinal Study on Aging, a large, national prospective study. Qualitative data on health, social participation and health equity were collected from focus group discussion with 120 participants in five Canadian communities. The authors listed the expected outcomes of the research, which included implication on age-friendly policies, services and structures to improve health, social participation and health equity in the communities.

Lewis, J. L., & Groh, A. (2016). It's about the people...: Seniors' perspectives on age-friendly communities. In T. Moulaert & S. Garon (Eds.), *Age-friendly cities and communities in international comparison: Political lessons, scientific avenues, and democratic issues* **(Vol. 14, pp. 81–98). Switzerland: Springer**.

Building upon the WHO Age-Friendly Cities and Communities framework, this paper reported on the empirical research to examine contextualised age-friendly initiatives in Waterloo, Canada. Through interviews with older adults on the liveability of Waterloo city for seniors, several themes centred on transportation, walkability, housing and social inclusion, were revealed. Findings revealed that both built and social environments were of importance to older adults. In addition, the relationship between housing and the surrounding neighbourhood was highlighted as a key concern to participants. Social respect and inclusion were identified as important elements to allow older adults to age in place.

**Menec, V. H., Brown, C. L., Newall, N. E.G., & Nowicki, S. (2016). How impor-
tant is having amenities within walking distance to middle-aged and older adults,
and does the perceived importance relate to walking?** *Journal of Aging and
Health, 28*(3), 546–567.

This paper explored the availability of amenities within walking distance to middle-
aged and older people, and its relationship with walking behaviour, activity level,
health outcomes and socio-demographic in a mid-western Canadian city. Logistic regression and ordinal least squares regression were applied on information
of neighbourhood amenities, socio-demographic, health, functional limitations collected from interviews and activity level measured by pedometers with 778 respondents aged between 45 and 94 years. The study demonstrated the complexity of walking behaviour to amenities. Majority of the participants did not consider amenities
within walking distance to be important; they were more likely to drive to amenities instead. Results suggested that self-reported walking to parks had a relationship
with overall activity. Although proximity to amenities was assumed as a strategy for
achieving age-friendly communities, the current study provided evidence that it did
not match the actual preferences of older adults albeit in a Canadian city. The author
recommended future studies on transportation patterns and barriers of walking to
amenities, which would encourage active healthy ageing lifestyle.

**Menec, V. H., Hutton, L., Newall, N., Nowicki, S., Spina, J., & Veselyuk, D. (2015).
How 'age-friendly' are rural communities and what community characteristics
are related to age-friendliness? The case of rural Manitoba, Canada.** *Ageing &
Society, 35*(1), 203–223.

This study evaluated age-friendliness across 56 communities of Age-Friendly Manitoba Initiative in mid-western Canada. Multi-level regression analysis was applied
on survey data of 7 domains of age-friendliness (physical environment; housing
options; social environment; opportunities for participation; community supports and
health-care services; transportation options; and communication and information)
with 1373 participants in 56 smaller and mostly rural communities across the Manitoba Province, and the 2006 census data of community characteristics (population
size; percentage of older adults aged 65 years and above; education; income). Results
indicated that overall age-friendliness, specifically individual component of social
environment, opportunities for participation, and communication and information,
was associated with higher proportion of older adults in the community. The lowest
age-friendliness was rated in small communities within a census metropolitan area
and remote communities in far north Manitoba, which emphasised the significance
of degrees of rurality. This study highlighted the importance of holistic consideration
of various community characteristics in the evaluation of age-friendliness.

**Menec, V. H., Means, R., Keating, N., Parkhurst, G., & Eales, J. (2011).
Conceptualizing age-friendly communities.** *Canadian Journal on Aging, 30*(3),
479–493.

Building upon the WHO Age-Friendly Cities framework, this paper conceptualised
the idea of age-friendliness through an ecological approach to provide direction for

future research and policy. Using 'social connection' as a heuristic metaphor, the paper proceeded to outline five principles from ecology literature that could help to elucidate the concept of age-friendliness. The first principle was interrelatedness of domains where a holistic approach spanning multiple domains should be adopted to consider complex urban issues. The second was the idea of relative proximity where environmental influences varied in their extent of influence on the older person. The third principle stressed the importance of considering individual characteristics of older people and not homogenising them when planning for design interventions. The fourth principle illustrated the dynamism of physical and social landscapes and highlighted the need to consider the implications of changes. The fifth principle concerned 'leverage points' that were key built environment domains that should be prioritised. The conceptual framework raised two key implications for older adults: (1) to ensure person-environment fit, older adults must play a participatory role in the planning and design of their living environments; and (2) to safeguard against the devolution of civic responsibilities that might arise from viewing age-friendliness as a way to offload such responsibilities onto the community.

Menec, V. H., Newall, N. E.G., & Nowicki, S. (2016). Assessing communities' age-friendliness: How congruent are subjective versus objective assessments? *Journal of Applied Gerontology, 35*(5), 549–565.

This paper assessed the congruence of age-friendliness assessments between subjective and objective measurements in Manitoba, Canada. Seven domains of age-friendliness were assessed subjectively and objectively through community resident survey with 990 individuals in 39 communities, and municipality survey with 130 respondents respectively. Results suggested that the subjective community resident survey was congruent with objective municipal survey even though age-friendliness measured by objective municipal measurement was relatively better than the subjective measurement. Using census data on socio-demographic factors, community characteristic of the predictors for age-friendliness were tested for both measurements and showed no significant difference. Although the authors acknowledged the municipal data as a reasonable source to assess age-friendliness in terms of cross-community comparisons, it might not be sufficient to reflect the perspective of residents.

Menec, V. H., Novek, S., Veselyuk, D., & McArthur, J. (2014). Lessons learned from a Canadian province-wide age-friendly initiative: The Age-Friendly Manitoba Initiative. *Journal of Aging & Social Policy, 26*(1–2), 33–51.

This paper discussed the Age-Friendly Manitoba Initiative and its enablers and barriers. Applying formative evaluation on the implementation process of the initiative from the perspective of five areas (i.e. organisation and management of the age-friendly initiative at the local level; community assessment; inter-sectoral partnerships; promoting age-friendliness; and projects implemented), 67 interviews were conducted in 44 rural and urban communities, which implemented at least one age-friendly project. This province-wide age-friendly initiative demonstrated considerable progress with Age-Friendly Committee formed by almost every community, and community assessments conducted to prioritise developments. Insufficient funding and capacity, specifically in small communities, and the lack of leadership or direction were the main barriers. In order to become more age-friendly, the findings emphasised the role of leadership, support towards communities and continuous local promotion at different scales.

Novek, S., & Menec, V. H. (2014). Older adults' perceptions of age-friendly communities in Canada: A photovoice study. *Ageing & Society, 34*(6), 1052–1072.

This study examined the perception of age-friendliness from the perspective of older adults in Manitoba, Canada. Photovoice technique, a participatory methodology of photography, interviews and group discussions, was used with 30 community dwelling older adults aged between 54 and 81 years in one urban community and three rural communities. According to theme analysis, priorities and barriers towards age-friendly communities were identified and grouped into three age-friendly features including age-friendly features, contextual factors and cross-cutting themes, which were generally consistent with the WHO domains of age-friendliness. Three contextual factors (i.e. community history and identity; ageing in urban, rural and remote communities; and environmental conditions) were explored that contributed to perception, experience and characteristics of age-friendly communities for older adults.

Orpana, H., Chawla, M., Gallagher, E., & Escaravage, E. (2016). Developing indicators for evaluation of age-friendly communities in Canada: process and results. *Health Promotion and Chronic Disease Prevention in Canada, 36*(10), 214–223.

This paper presented the process of indicator identification for age-friendly community initiatives evaluation to address the collective impact and cross-comparison among Canadian communities. Drawing on literature review and the WHO Age-friendly Cities framework, two iterative consultations with 240 stakeholders generated 39 indicators in the 8 age-friendly city domains, and 4 indicators for long-term health and social outcomes. Available as a user-friendly guide and as the last milestones of the Pan-Canadian age-friendly city initiative, the age-friendly city indicators offered a potential tool for age-friendly city evaluation within communities to further improve these initiatives.

Spina, J., Smith, G. C., & DeVerteuil, G. P. (2013). The relationship between place ties and moves to small regional retirement communities on the Canadian prairies. *Geoforum, 45*, 230–239.

This research explored place ties and their impact on older adult's decision on migration to regional retirement communities in Manitoba, Canada. Quantitative descriptive and qualitative analysis were applied on information collected from a two-stage survey with 34 older migrants aged 55 years and older who moved to retirement communities. Results suggested place ties associated destination community was the leading factor on relocation decision-making of older adults. Previous place experience and individual social contacts also played an important role in migration as well as population growth, community sustainability and regional competitiveness.

Syed, M. A., McDonald, L., Smirle, C., Lau, K., Mirza, R. M., & Hitzig, S. L. (2017). Social isolation in Chinese older adults: Scoping review for age-friendly community planning. *Canadian Journal on Aging, 36*(2), 223–245.

This review aimed to further understand social isolation and loneliness among Canadian and other Western urban-dwelling Chinese older people. Drawing on the WHO Age-friendly Cities framework and 8 key domains, a six-step scoping review was undertaken that screened 19 texts (Canada 10, United States 8, Australia 1). Results revealed the issue of social isolation and loneliness among Chinese older adults, and all 8 domains were discussed to some extent in the literature, particularly social participation, community support and health services, housing, and communication and information. Although the findings provided evidence that ethnic-minority urban-dwelling older adults might experience less social isolation and loneliness through age-friendly strategies (e.g. multi-sectoral interventions with multiple stakeholders) in Western communities, the authors recommended further research across communities with various methodology design.

Winters, M., Barnes, R., Venners, S., Ste-Marie, N., McKay, H., Sims-Gould, J., & Ashe, M. C. (2015). Older adults' outdoor walking and the built environment: Does income matter? *BMC Public Health, 15*(1), 876–883.

This study addressed the research question—whether walkability affected outdoor walking among older Canadians, and if this was moderated by socio-economic status. Data on objective walkability and outdoor walking were assessed by Street Smart Walk Score® and self-reported timing of outdoor walking separately from 1309 older adults aged 65 and above in the Canadian Community Health Survey Healthy-Aging 2008–2009 Cycle. Findings from multivariate logistic regression showed that higher score on walkability had significant association with outdoor walking of more than 150-minute moderate to vigorous activity per week (physical activity guidelines). Older adults living in neighbourhoods identified as 'walker's paradise' were three times or more likely to meet physical activity guidelines than those living in neighbourhoods categorised as 'car-dependent/very car dependent'. However, no moderating effect was found between walkability and outdoor walking through household income. The authors concluded that neighbourhood design might be a strategy to encourage older adults to walk more and thus, improve their health.

Winters, M., Voss, C., Ashe, M. C., Gutteridge, K., McKay, H., & Sims-Gould, J. (2015). Where do they go and how do they get there? Older adults' travel behaviour in a highly walkable environment. *Social Science & Medicine, 133*, 304–312.

This study assessed the mobility of Canadian older adults dwelling in the most walkable neighbourhood environments. Information on general health, physical activity, neighbourhood environment and social connections, and data of seven-day travel diary were obtained through survey and tri-axial accelerometers separately among older adults (aged 60 years and older) living in highly walkable Vancouver's downtown core. Descriptive statistics showed grocery stores, restaurants, malls/marketplaces, and others' homes were the key destinations for older adults, and their average travel-related mobility and physical activity-related mobility were 4.6 one-way trips/day with the majority by walking (62.8%) or cycling (3.2%) and 7910.1 steps/day with 39.2 min/day of moderate to vigorous physical activity. Multivariate linear regression models found that daily trip frequency had significant positive relationship with physical activity outcomes. Findings suggested that age-friendly neighbourhood (i.e. walkable environment and accessible to multiple destinations) played an important role in older adults' travel- and physical-related activities, which in turn might promote health among older adults.

2.5 Europe

2.5.1 United Kingdom

Bowling, A., & Stafford, M. (2007). How do objective and subjective assessments of neighbourhood influence social and physical functioning in older age? Findings from a British survey of ageing. *Social Science & Medicine, 64*(12), **2533–2549.**

This paper examined the relationship between the objective and subjective neighbourhood characteristics and older people's social and physical functioning. More specifically, it studied the associations between socio-economic characteristics of neighbourhoods, older people's neighbourhood perceptions and older people's social and physical functioning. The study aimed to find out whether the type of neighbourhoods impacted social and physical functioning of older adults. Using multi-level modelling, the responses from 786 older adults aged 65 years and older were analysed. The results—independent of individual demographic and socio-economic characteristics—suggested that: (1) the respondents who lived in less affluent neighbourhoods were more likely to have lower levels of social activities; (2) perceiving the neighbourhood as less neighbourly was associated with lower levels of social activities; and (3) perceiving neighbourhood facilities negatively was associated with lower levels of social activities.

Brookfield, K., Ward Thompson, C., & Scott, I. (2017). The uncommon impact of common environmental details on walking in older adults. *International Journal of Environmental Research and Public Health, 14*(2).

Informed by the ecological model of person-environment fit, this qualitative study examined the relationship between the environment and older adults' walking behaviour. Some 22 purposively sampled older adults aged 60 years and above from Edinburgh, Scotland, participated in semi-structured interviews and focus group discussions on age-, dementia- and stroke-friendly environments. Results showed that the environment coupled with person-related factors such as functional and cognitive impairments influenced the walking behaviour of older adults. Older adults identified uneven and tactile pavements and curbs as barriers to walking whereas benches, toilets and handrails were seen as enablers. The study also provided insights into how physical aspects of the built environment informed older adults' walking behaviour but also their perceptions such as the assumptions, meanings and expectations.

Buffel, T., Skyrme, J., & Phillipson, C. (2017). Connecting research with social responsibility: Developing 'age-friendly' communities in Manchester, UK. In D. T. L. Shek & R. M. Hollister (Eds.), University Social Responsibility and Quality of Life (Vol. 8, pp. 99–120). Springer.

This chapter examined the goals of the University of Manchester's social responsibility through analysing a research project's contribution to social responsibility. The research project was about neighbourhood perception and active ageing. It was chosen by the World Health Organization as the best practice example of involving older people as co-investigators in researching and developing age-friendly cities. The authors examined the project's aims, methods, research activities and process, and the involvement of older adults as co-investigators. One of the main hypotheses was that age-friendly initiatives would not work without the co-production method or active involvement of older adults.

Burton, E. J., Mitchell, L., & Stride, C. B. (2011). Good places for ageing in place: Development of objective built environment measures for investigating links with older people's wellbeing. *BMC Public Health*, *839*(11).

This paper reported on the development and testing of the Neighbourhood Design Characteristics Checklist (NeDeCC), a tool that objectively assessed the built environment characteristics. The tool was developed through a review of urban design literature, design guidance documents and existing measures. It was tested in a study with 200 older adults aged 65 years and older from Oxfordshire, Gloucestershire and Greater Manchester, UK, who participated through in-depth interviews. The interviews measured place-related wellbeing and were complemented with the assessment of the neighbourhood environment using NeDeCC. Findings showed that most of the built environment measures were significantly related to the housing type of older adults, and that there was a significant relationship between wellbeing and several of the built environment characteristics. However, because of the exploratory nature of the study, the relationships were unclear.

Liddle, J., Scharf, T., Bartlam, B., Bernard, M., & Sim, J. (2013). Exploring the age-friendliness of purpose-built retirement communities: Evidence from England. *Ageing and Society*, *34*(09), 1601–1629.

This paper reported on the longitudinal study of purpose-built retirement community's age-friendliness. The paper proposed a new definition of age-friendly community that perceived age-friendliness as an ongoing process. A mixed-method approach was used to gather the data. Two surveys were conducted with 122 and 156 older residents in 2007 and 2009 respectively. Additionally, questionnaires were given to staff members in 2010; semi-structured interviews were conducted with those who were involved in the redevelopment and management of the retirement community, and two focus group discussions were conducted with staff members. The results showed that, among other things, the studied retirement community could become an age-friendly community through improving ageing in place for the residents, addressing age segregation, improving pavements outside the village and providing accessible transport.

McGarry, P. (2018). Developing age-friendly policies for cities: Strategies, challenges and reflections. In Buffel T., Handler S., & Phillipson C. (Eds.), *Age-friendly cities and communities: A global perspective* (pp. 231–250). Bristol: Bristol University Press.

This chapter examined United Kingdom's ageing policies and strategies since the 1990s, in particular, the development of Age-Friendly Manchester. Began in the early 1990s, Manchester gradually adopted an asset-based account of ageing, formed a Better Government for Older People Group that encouraged the engagement of older adults in the community, the Valuing Older People Partnership that produced several age-friendly initiatives, and finally Age-friendly Manchester in 2010. The author discussed how the global financial crisis and economic austerity impacted on Age-friendly Manchester and looked at the prospects of developing Greater Manchester region into UK's first age-friendly city region and global centre of excellence for ageing.

Murray, A., & Musselwhite, C. (2019). Older peoples' experiences of informal support after giving up driving. *Research in Transportation Business & Management, 100367*.

This paper explored retired drivers' use of informal support after giving up driving. Through semi-structured interviews with seven older adults, multifaceted domains inclusive of practical and psychosocial factors of informal support were identified. These older adults more frequently pursued informal support due to ill health and living away from bus services. Findings revealed that informal support functioned at a community scale and was highly dependent on social capital, suggesting a need for improved understanding to better activate it.

Peace, S., Katz, J., Holland, C., & Jones, R. (2018). The age-friendly community: A test for inclusivity. In Buffel T., Handler S., & Phillipson C. (Eds.), *Age-friendly cities and communities: A global perspective* (pp. 251–272). Bristol: Bristol University Press.

This chapter examined and tested the inclusivity of older people with sight loss and living in English urban and rural communities. Drawing on two studies—age-friendly neighbourhood programme project in Manchester and the Needs and Aspirations of Vision Impaired Older People (NAVIOP) study—the authors used data of 50 older people with visual impairment interviewed in the NAVIOP study to examine the meaning of age-friendliness. Housing, home activities, wider community involvement and physical environment were discussed. The most important issues for older adults with visual impairment were adapted housing, accessibility, familiarity of the outdoor environment and social connectedness.

Thompson, C. W., Curl, A., Aspinall, P., Alves, S., & Zuin, A. (2014). Do changes to the local street environment alter behaviour and quality of life of older adults? The 'DIY Streets' intervention. *British Journal of Sports Medicine*, 48(13), 1059–1065.

This longitudinal study examined the associations between residential street improvements in the UK and older adults' outdoor activities, health and quality of life. The study employed social ecological models of behaviour and conducted three types of surveys: pre- and post-intervention cross-sectional survey, longitudinal cohort survey, and activity survey. Older adults aged 65 years and older were recruited for the cross-sectional survey and invited to participate in the longitudinal and activity study. Results showed that street improvements influenced older adults' perceptions of street walkability and safety at night, but not activity levels, health or quality of life.

White, S., & Hammond, M. (2018). From representation to active ageing in a Manchester neighbourhood: Designing the age-friendly city. In T. Buffel, S. Handler, & C. Phillipson (Eds.), *Age-friendly cities and communities: A global perspective* (pp. 193–210). Bristol: Bristol University Press.

This chapter explored the capability approach to age-friendly design and reported on a community-led research project that was conducted in Manchester, UK in 2012. The project focused on the lived experience of older adults and explored the applicability of Manchester's age-friendly city programme in the Old Moat neighbourhood. Following the capability approach, residents and institutional stakeholders worked together on a neighbourhood action plan to make the neighbourhood more age-friendly. The project demonstrated that age-friendly design was both a participatory/collaborative and spatial endeavour.

2.5.2   Other European Countries

Camporeale, R., Wretstrand, A., & Andersson, M. (2019). How the built environment and the railway network can affect the mobility of older people: Analyses of the southern Swedish region of Scania. *Research in Transportation Business & Management, 100368*.

This paper examined the built environment and the railway network's influence on the spatial distribution and mobility of older adults aged 65 years and older in Scania, a region in Sweden. The study used longitudinal datasets for health insurance and labour market, travel patterns and travel behaviour, and property-specific data. It provided an overview of older population distribution, the number of older residents in urban areas, and residential mobility across these areas. The analysis included an investigation of an association between older adults' residential mobility and socio-demographic characteristics and travel survey variables. Results showed the spatial unevenness of ageing in the Scania region, high car dependence among older adults, and how the expansion of the rail service might contribute to improved mobility, among other things.

Domínguez-Párraga, L. (2019). Neighborhood Influence: A qualitative study in Cáceres, an aspiring age-friendly city. *Social Sciences, 8*, 195.

This qualitative study examined older adults' perceptions of two selected neighbour-hoods and their composition in Cáceres, Spain. The theoretical underpinnings of the study comprised sociology of ageing (activity theory of ageing) and urban sociology. A total of 32 in-depth interviews with older adults aged 65 years and above were conducted. Interviewees were asked about their activities, social relationships, satisfaction with neighbourhood, and perceived health and wellbeing. The analysis was based on grounded theory. Results showed that the social aspects of neighbourhood life played the most important role in determining older adults' perceptions of their neighbourhoods.

Dryjanska, L., & Giua, R. (2019). The past empowering the present: Intergenerational solidarity improving the quality of life. In A. Bianco, P. Conigliaro, & M. Gnaldi (Eds.), *Italian Studies on Quality of Life* (pp. 255–272). Springer.

This chapter examined intergenerational solidarity and quality of life in Rome, Italy. The authors studied two types of intergenerational activities, one involving older adults and young adults and the other involved older adults and children. The study used the reminiscence approach and qualitative participant observation. The group of older adults and young adults produced a documentary on memories of Rome while the group of older adults and children participated in a Christmas-related programme that involved reading, singing, etc. Results showed that given the cultural context of Rome, the intergenerational activities proved to be successful. Where conflicts arose, they were addressed in terms of problem solving by all participants.

Green, G. (2013). Age-friendly cities of Europe. *Journal of Urban Health*, *90*(1), 116–128.

This paper described and assessed the application of the WHO healthy ageing approach to policy and programme development in 59 European cities that were in the European Healthy Cities Network. The data was sourced from the General Evaluation Questionnaire, the network's Annual Reporting Templates, and reports of the seven Sub-Network meetings during Phase IV. The key messages of this analysis were that cities adopted the healthy ageing approach and focused on enabling older adults rather than on the limitations that age imposed on people; adopting the WHO healthy ageing approach might contribute to diminishing decrepitude and dependence and enhance older adults' independence and achievement.

Haacke, H. C., Enßle, F., Haase, D., Helbrecht, I., & Lakes, T. (2019). Why do(n't) people move when they get older? Estimating the willingness to relocate in diverse ageing cities. *Urban Planning, 4*(2).

This paper examined the association between older adults' personal attributes (such as social class, gender, age, migrant history) and their planned and actual relocation in Germany. The aim was to identify and understand the factors affecting residential movement in later life as the immediate neighbourhood played an important role in the older person's social integration. A quantitative survey was conducted with 427 older adults aged 60 years and older in Berlin, Germany. Another 18 expert interviews as well as focus group discussions with 26 older adults were conducted. The authors used descriptive statistics, chi square test, one-way ANOVA and multivariate binomial logistic regression to understand the relationship between the variables. Results showed that age impacted older people's past or planned movement, especially among those aged between 65 and 75 years.

Jackisch, J., Zamaro, G., Green, G., & Huber, M. (2015). Is a healthy city also an age-friendly city? *Health Promotion International, 30*(Suppl 1), i108–i117.

This paper analysed empirical data of cities from the WHO European Healthy Cities Network, their implementation of the concept and practice of age-friendly environments between 2009 and 2013. Both qualitative and quantitative evidence were evaluated using the realist synthesis methodology. The authors reviewed 33 case studies from 32 cities including the General Evaluation Questionnaire submitted by 71 cities, and other supplementary data such as Annual Reporting Template of the Network cities, etc. The authors mapped activities across age-friendly domains and identified the interventions that made cities more age-friendly: removing barriers and creating supportive physical environments, creating resilient social environments, and providing health and social services in the community. The analysis also revealed that political commitment was necessary to realise these initiatives.

Lager, D., Van Hoven, B., & Huigen, P. P. (2016). Rhythms, ageing and neighbourhoods. *Environment and Planning A: Economy and Space, 48*(8), 1565–1580.

This study used Lefebvre's Rhythmanalysis to investigate the everyday temporal rhythmic ordering of people and place in 53 older adults (aged 65 years and older) in Groningen, the Netherlands as a way to understand the experiential dimension of ageing in urban neighbourhoods. Results showed how the older person embodied activity in their everyday lives in terms of how they valued their own rhythms, e.g. slowing rhythms were coupled with an increased need for rest time. The authors suggested that the lack of synchronised rhythms—time and place, e.g. between the young and old could result in 'generational divide' experiences.

McDonald, B., Scharf, T., & Walsh, K. (2018). Creating an age-friendly county in Ireland: Stakeholders' perspectives on implementation. In T. Buffel, S. Handler, & C. Phillipson (Eds.), *Age-friendly cities and communities: A global perspective* **(pp. 143–166). Bristol: Bristol University Press.**

This chapter charted the evolution of Ireland's Age Friendly Cities and Counties Programme and analysed the development and implementation of a local age-friendly initiative—Fingal's Age Friendly County initiative. Older adults and stakeholders were interviewed. The data was analysed to answer two questions: how stakeholders influenced the development of the initiative, and how they influenced the activities that involved older people. Findings showed that despite austerity, Ireland's programme was quite successful. Strong political leadership, policy framework, individual and organisational leadership were critical success factors.

McGarry, P. (2012). Good places to GrowOld: Age-friendly cities in Europe. *Journal of Intergenerational Relationships*, *10*(2), 201–204.

In reviewing the European age-friendly cities movement, the author argued for cities to consider older people and a linked up, programmatic response when developing policies and programmes for age-friendly cities and communities. Examples of key partnerships and action areas were shared.

Moulaert, T., & Houioux, G. (2016). A Belgian case study: Lack of age-friendly cities and communities knowledge and social participation. In T. Moulaert & S. Garon (Eds.), *Age-friendly cities and communities in international comparison: Political lessons, scientific avenues, and democratic issues* **(Vol. 14, pp. 213–229). Cham: Springer.**

This chapter examined the WHO Age-friendly Cities model and how local actors in the Walloon Region, Belgium, could use it. The study used a qualitative exploratory study design with 12 municipalities where interviews were conducted with three types of actors within each city—elected politicians, administrative staff and older adults. Results showed that field workers and politicians had poor understanding of the age-friendly model and the needs of older adults. The authors suggested 'participatory diagnosis' as a potentially good strategy and method to take older adults' needs into consideration when developing age-friendly community.

Parent, A. S., & Wadoux, J. (2016). Toward an age-friendly European Union: An interview In T. Moulaert & S. Garon (Eds.), *Age-friendly cities and communities in international comparison: Political lessons, scientific avenues, and democratic issues* **(pp. 247–259). Cham: Springer.**

This chapter contained an interview with Anne-Sophie Parent and Julia Wadoux, AGE Platform Europe (AGE). AGE was a European platform that consisted of 150 organisations, representing more than 40 million older adults in the European Union (EU). The interviewees spoke about AGE's mission to promote age-friendly environments in the EU; their cooperation with the World Health Organisation; their project on Age-Friendly Environments Innovation Network (AFE-INNOVNET); and the EU Covenant on Demographic Change and its relation to the WHO Global Network for Age-friendly Cities and Communities.

Portegijs, E. and Rantanen, T. (2019). Life-space mobility and active ageing. In A. P. Lane (Ed.), *Urban environments for healthy ageing: A global perspective*. Routledge.

This chapter examined the life-space of older adults and active ageing using data from the research project, Life-Space Mobility in Old Age where 848 Finnish older adults aged 75 years and older in Central Finland were studied and reassessed annually for two years. Findings showed that the decline in life-space mobility was associated with the development of another disability in activities of daily living and lower quality of life. The authors also reported on the active ageing intervention study, Volunteering, Access to Outdoor Activities and Wellbeing in Older People, which examined the needs of older adults aged 65 years and older with severe mobility limitations. Older adults and volunteers took part in several activities that expanded participants' life-space. Findings showed that going out with a volunteer helped the participants to perceive the outdoor spaces as more accessible. The authors concluded that it was important for societies to promote active ageing among older adults.

Ribeiro, A. I., Mitchell, R., Carvalho, M. S., & de Pina Mde, F. (2013). Physical activity-friendly neighbourhood among older adults from a medium size urban setting in Southern Europe. *Preventive Medicine*, *57*(5), 664–670.

This paper reported on the cross-sectional study that examined the associations between neighbourhood socio-environmental characteristics and the frequency of leisure-time physical activity (LTPA) among older adults in Porto, Portugal. Some 580 older adults aged 65 years and above were studied using Geographic Information System and Generalized Additive Models to check for associations between the objective neighbourhood characteristics and LTPA of older adults. The analysis showed no relationship between the two. However, neighbourhood characteristics seemed to influence LTPA of already-active persons, i.e. those older adults who were already engaged in some sort of physical activity.

Rogelj, V., & Bogataj, D. (2018). Planning the home and facility-based care dynamics using the multiple decrement approach: The case study for Slovenia. *IFAC-PapersOnLine*, *51*(11), 1004–1009.

This paper interrogated the demand and dynamics of long-term care premium for age-friendly healthcare facilities and services in Slovenia. The authors proposed a spatial interaction model to optimise the appropriation of resources for care, logistics and housing by forecasting the needs of different age-cohorts based on declining functional abilities. In recognising the heterogeneity of different age-cohorts, insurers could gain greater precision in calculating contribution rates and insurance premiums over the lifetime horizon of older adults, potentially reducing the cost of care.

Sánchez-González, D., & Rodríguez-Rodríguez, V. (2016). *Environmental Gerontology in Europe and Latin America* **(Vol. 13). Springer**.

This book (15 chapters) explored the relationships between the physical-social environment and quality of life of older people in Europe and Latin America using environmental gerontology, geographical and psycho-social approaches and socio-spatial analysis. A range of environments—housing, public spaces, landscapes, neighbourhoods, urban and rural areas—were analysed at different scales—macro (e.g. urban and rural environments), meso (e.g. neighbourhood, public space) and micro (e.g. personal, home, institution). The book discussed objective and experiential person-environment transactions as people aged, issues of place attachment and health as well as the challenges and problems of age-friendly environments with the use of case studies, e.g. contrasting Manchester, Ghent and Brussels. Latin American countries covered included Mexico, Chile, Ecuador and Brazil.

Simone, P., & Le Borgne-Uguen, F. (2016). "Age-friendly cities" in France: Between local dynamics and segmented old age policy. In T. Moulaert & S. Garon (Eds.), *Age-friendly cities and communities in international comparison: Political lessons, scientific avenues, and democratic issues* **(Vol. 14, pp. 191–213). Springer**.

This chapter examined the WHO Age-Friendly Cities approach in France that was incorporated in its national programme on ageing, 'Aging Well'. The study looked at three cities in France and their local, ageing-related policies. The chapter also compared the WHO and the French ageing programmes, finding several differences between the two in the three cities. In particular, the findings revealed that the segmentation of local policies diminished the effectiveness of the age-friendly city approach and that the understanding of ageing in France revolved around both physical environment and citizenship.

Štaube, T., Leemeijer, B., Geipele, S., Kauškale, L., Geipele, I., & Jansen, J. (2016). Economic and financial rationale for age-friendly housing. *Journal of Financial Management of Property and Construction*, *21*(2), 99–121.

This study conducted review of primary and secondary literary sources, logical approach and comparison of real estate and socio-economic regional statistics on the socio-economic aspects of age-friendly housing cases in two countries, the Netherlands and Latvia. The authors argued that age-friendly housing construction would become the standard of sustainable property development, and a thorough calculation and understanding of its financial viability was required.

Van Cauwenberg, J., De Donder, L., Clarys, P., De Bourdeaudhuij, I., Buffel, T., De Witte, N., Dury, S., Verte, D., & Deforche, B. (2014). Relationships between the perceived neighborhood social environment and walking for transportation among older adults. *Social Science & Medicine*, *104*, 23–30.

This cross-sectional study employed an ecological model to examine the associations between older adults' perceived social environment and their walking for transportation in Belgium. Between 2004 and 2010, 50,986 Flemish older adults aged

65 years and older took part in a survey on older adults' walking for transportation, individual perceived physical and social environmental factors. Multilevel logistic regression analyses were conducted and showed significant positive relationships between walking for transportation and the social neighbourhood environment, suggesting that projects stimulating interpersonal relationships, place attachment and formal community engagement might promote walking for transportation among older people.

Van Dijk, H. M., Cramm, J. M., Van Exel, J. O. B., & Nieboer, A. P. (2014). The ideal neighbourhood for ageing in place as perceived by frail and non-frail community-dwelling older people. *Ageing and Society, 35*(08), 1771–1795.

This study examined older adults' perceptions of neighbourhood physical and social characteristics that supported ageing in place in Rotterdam, the Netherlands. Using the WHO Global Age-friendly Cities guide as a theoretical framework and Q-methodology, the study developed 26 statements about ideal neighbourhood characteristics for rank-ordering by 16 frail and 16 non-frail older adults aged 70 years and older. The analysis showed an emphasis on the importance of maintaining independence and the WHO key domains of outdoor spaces and buildings, transportation, housing, and community support and health appeared to be most essential to older people. The authors concluded that older people's dependence on the neighbourhood was not static and was determined by levels of frailty and changing social and physical neighbourhood conditions.

Van Hees, S., Horstman, K., Jansen, M., & Ruwaard, D. (2017). Photovoicing the neighbourhood: Understanding the situated meaning of intangible places for ageing-in-place. *Health & Place, 48*, 11–19.

This paper examined the meaning and construction of ageing in place in the development of lifecycle-robust neighbourhoods in the Netherlands. Using the constructivist and photovoice approach, 14 professionals and 18 older adults aged 70 years and above were asked to take photographs of places that enabled ageing in place and discuss their experiences. The results showed that while professionals attached the most importance to the physical built environment, older adults emphasised historical characteristics of places and other meaningful, intangible neighbourhood characteristics.

Van Hees, S., Horstman, K., Jansen, M., & Ruwaard, D. (2018). Meanings of 'lifecycle robust neighbourhoods': Constructing versus attaching to places. *Ageing and Society, 38*(6), 1148–1173.

This paper examined the meaning of 'lifecycle-robust neighbourhoods' in relation to the concept of ageing in place and age-friendly communities through the case study of Voor Elkaar in Parkstad, the Netherlands. Voor Elkaar was developed as a lifecycle-robust neighbourhood in 2011 to enable older people to live independently for longer. Ethnographic methods were used—interviews, observations and focus groups—with 28 developers (e.g. policymakers, representatives of older people) and 28 users (older people who lived independently) to examine their experiences of and thoughts on ageing, their homes and neighbourhoods, and understand the meanings they gave to ageing and lifecycle-robustness. Results showed that the idea of lifecycle robust neighbourhood resonated with the ideals of age-friendly places. As to be expected, perspectives differed with developers often talking in terms of constructing places, e.g. about enabling and disabling elements while older people as users did not construct but experienced places while living and ageing there.

van Hoof, J., Kazak, J. K., Perek-Bialas, J. M., & Peek, S. T. M. (2018). The challenges of urban ageing: Making cities age-friendly in Europe. *International Journal of Environmental Research and Public Health, 15*(11).

This viewpoint article examined the challenges of creating age-friendly cities in Europe. More specifically, the study focused on inclusive neighbourhoods and technology as an enabler of ageing in place in two European cities, the Hague in the Netherlands (longstanding reputation for age-friendly measures) and Cracow in Poland (one of the youngest countries in Europe but would soon be experiencing rapid population ageing). Various age-friendly projects in the two cities were described and examined against the 8 WHO domains. The analysis showed that the Hague's main challenges were with improving the vitality of older adults, loneliness and ageing in place. Cracow's main challenges were on improving education of older population, active and healthy ageing, intergenerational integration, and participation in social activities.

van Hoof, J., Dikken, J., Buttiġieġ, S. C., van den Hoven, R. F. M., Kroon, E., & Marston, H. R. (2019). Age-friendly cities in the Netherlands: An explorative study of facilitators and hindrances in the built environment and ageism in design. *Indoor and Built Environment.* https://doi.org/10.1177/ 1420326X19857216.

This paper investigated the age-friendly city domains and the presence of ageism in the Dutch municipalities of the Hague and Zoetermee. The analysis was done on a neighbourhood level. Researchers used a photoproduction method and took 620 photographs of neighbourhoods following the WHO Checklist of Essential Features of Age-Friendly Cities. Implicit and explicit ageism was analysed based on the photographs. In both municipalities, facilitators and hindrances in the built environment were identified, especially in the domains of Communication and information, Housing, Transportation, Community support and health services, and Outdoor spaces and buildings. Instances of both positive and negative ageism were predominantly found in the domain of Communication and Information.

Walker, A. (2016). Population ageing from a global and theoretical perspective: European lessons on active ageing. In T. Moulaert & S. Garon (Eds.), *Age-friendly cities and communities in international comparison: Political lessons, scientific avenues, and democratic issues* **(Vol. 14, pp. 47–64). Switzerland: Springer.**

Building upon the WHO idea of 'active-ageing', the chapter argued that the lack of preciseness of what constituted 'active-ageing' was impeding its effective implementation as a policy strategy in Europe, the world's oldest region. Barriers to active ageing, ranging from political to cultural, societal, bureaucratic and unequal ageing were identified. Recommendations included clarifying active ageing and age-friendly action, providing a comprehensive strategy to active ageing and incorporating clear and consistent principles to ensure that the initiatives were comprehensive, consistent and responsive to the needs of older adults. The author concluded with an affirmation that active ageing and age-friendly cities policies should be integrated—its integration 'could have the potential to transform the ageing experience globally'.

2.6 General (Other Regional Works)

Aboderin, I., Kano, M., & Owii, H. A. (2017). Toward "age-friendly slums"? Health challenges of older slum dwellers in Nairobi and the applicability of the age-friendly city approach. *International Journal of Environmental Research and Public Health, 14*(10).

This paper examined the health-related challenges that a slum-focused age-friendly cities initiative in Sub-Saharan Africa would need to address, and the extent to which the WHO Age-friendly Cities framework offered an appropriate basis to do so. The study focused on two slum communities in Nairobi, Kenya—Korogocho and Viwandani. Existing qualitative and quantitative data on the two slums (context, health, etc.) and the results of the 2015 AFC Indicator Guide pilot study, which assessed the age-friendliness of the two slums, were examined. Results pointed to an adverse built environment with a lack of basic amenities, economic insecurity of older adults, and the absence of health services. Additionally, older adults seemed to experience mental and emotional stress, there were disparities in opportunities and wellbeing among older adults, and limited participation in decision-making. The indication was that there was scope to adapt the WHO domains to better fit the slum context.

Alhamdan, A. A., Alshammari, S. A., Al-Amoud, M. M., Hameed, T. A., Al-Muammar, M. N., Bindawas, S. M., Al-Orf, S. M., Mohamed, A. G., Al-Ghamdi, E. A., & Calder, P. C. (2015). Evaluation of health care services provided for older adults in primary health care centers and its internal environment. A step towards age-friendly health centers. *Saudi Medical Journal, 36*(9), 1091–1096.

This study evaluated the healthcare services for older adults in the primary health care centres in Riyadh, Kingdom of Saudi Arabia. A secondary aim was to recommend to policymakers how to make primary health care centres more age-friendly and effective. The authors used the WHO Age-friendly Primary Health Care Centres Toolkit to evaluate 15 primary health care centres across the city in terms of their services, screening protocols and environment. Findings were analysed and organised according to the evaluated themes: clinical services and health assessments, counselling services, accessibility, and signage. Results showed that certain improvements could be made, especially in cancer screening, vaccinations, annual comprehensive geriatric screening programmes, Braille signage, better public transport access and parking.

Bastani, P., Marzaleh, M. A., Dehghani, M., Falahatzadeh, M., Rahmati, E., & Tahernezhad, A. (2017). The status of Iranian hospital pharmacies according to age-friendly pharmacies criteria. *Journal of Advanced Pharmaceutical Technology & Research, 8*(4), 120–124.

This study examined the age-friendliness of 67 public and private hospital pharmacies in three cities (Shiraz, Mashhad and Isfahan) in Iran. The evaluation was done with a researcher-made checklist, which included questions on physical environment, healthcare, supply and logistics, and emotional aspects. Kruskal–Wallis and Mann–Whitney statistical tests were applied. No association was found between the type of pharmacies and the final score of age-friendliness. Only two pharmacies exhibited high levels of age-friendliness. The emotional aspect ranked the highest in terms of age-friendliness in all three cities, suggesting that it was important for older adults to be treated with respect in pharmacies. Certain improvements could be made in the areas of physical environment, supply and procurement of medicines in state pharmacies, and pharmaceutical supply and logistic services in private pharmacies.

Beard, J. R., & Montawi, B. (2015). Age and the environment: The global movement towards age-friendly cities and communities. *Journal of Social Work Practice, 29*(1), 5–11.

This paper outlined the development of a global movement towards the creation of age-friendly environments. The authors highlighted innovative age-friendly initiatives from member cities (New York, Akita, New Delhi and the City of Edmonton) in the WHO Global Network of Age-Friendly Cities and Communities.

Biggs, S., & Haapala, I. (2015). Age-friendly environments. *Journal of Social Work Practice, 29*(1), 1–3.

Building upon the WHO Age-Friendly Environments initiative, this editorial examined urban developments where older adults were active participants in age-friendly initiatives. Examples of urban case studies were cited from China, the Netherlands, Australia, Finland, Belgium, Canada (Quebec), Hong Kong and Switzerland. The compilation highlighted the contextual importance of space and its relationship with multigenerational interaction and the development of age-friendly initiatives.

Buffel, T., Handler, S., & Phillipson, C. (2018). Age-friendly cities and communities: A manifesto for change. In T. Buffel, S. Handler, & C. Phillipson (Eds.), *Age-friendly Cities and Communities* **(1 ed., pp. 273–288): Bristol University Press.**

This book outlined various guiding considerations for age-friendly policies. The main points included: (1) emphasis on context-specific approaches over generic approaches to tackle social inequality and exclusion, (2) creation of multi-stakeholder partnership, (3) diversity of housing services and amenities, (4) neighbourhood safety and security, and (5) participatory planning through co-ownership and community empowerment. The paper concluded with an evaluation of the efficacy of the WHO Global Network of Age-Friendly Cities and Communities and proposed recommendations to address future challenges.

Buffel, T., McGarry, P., Phillipson, C., De Donder, L., Dury, S., De Witte, N., Smetcoren, A. S., & Verte, D. (2014). Developing age-friendly cities: Case studies from Brussels and Manchester and implications for policy and practice. *Journal of Aging & Social Policy, 26*(1–2), 52–72.

This comparative study examined the policies and initiatives for active ageing in Brussels, Belgium, and Manchester, UK. The cities' cultural and socio-political background, and participation in the WHO Global Network of Age-Friendly Cities and Communities were reviewed. The similarities, differences, opportunities and barriers of the cities' age-friendly approach were analysed. Results showed that both cities used a framework for their age-friendly policies that was similar to the WHO global network; both cities provided opportunities for volunteering and other social activities of older adults; both cities used different branding of age-friendliness, e.g. Manchester had a stronger tradition in endorsing age-friendly policies that were also more culturally sensitive than those in Brussels.

Buffel, T., Phillipson, C., & Scharf, T. (2012). Ageing in urban environments: Developing 'age-friendly' cities. *Critical Social Policy, 32*(4), 597–617.

Against the backdrop of a myriad of age-friendly models, this paper argued that an evaluative approach was more suited over a definitive approach to understanding age-friendly cities. The paper highlighted poverty, urban hazards, social isolation and high crime rates as barriers to healthy ageing for older adults. Accessibility to public amenities and strong social support were critical to creating opportunities for healthy ageing amongst older adults. Recommendations included greater involvement of older adults in the socio-economic aspects of urban development.

Buffel, T., & Phillipson, C. (2016). Can global cities be 'age-friendly cities'? Urban development and ageing populations. *Cities, 55*, 94–100.

This paper outlined the development of various age-friendly models, with emphasis on the WHO Global Network of Age-Friendly Cities framework. This was followed by an analysis of the socio-economic pressures and their influence on the development of age-friendly urban environments, and exploration of barriers for improving age-friendly environments from the critical perspective of urban sociology. Results revealed that major challenges of applying age-friendly policies had arisen from financial cuts on social programmes, pressures from urban development, and privatisation of public spaces. The paper concluded with recommendations and stressed the importance of ensuring spatial justice for all user groups.

Buffel, T., & Phillipson, C. (2018). A Manifesto for the age-friendly movement: Developing a new urban agenda. *Journal of Aging & Social Policy, 30*(2), 173–192.

While the WHO Global Network of Age-Friendly Cities and Communities initiative had rallied support for age-friendly policies, this paper argued that challenges still existed towards successful implementation of age-friendly policies. These challenges included rising social inequality and the constraints of financial support for age-friendly policies. To address these challenges, it was recommended that the current paradigm of age-friendly policies be expanded to address social inequality, promote more inclusive participation, encourage funding through multidisciplinary collaborations and provide more opportunities to integrate research findings with policy implementation.

Cerin, E., Nathan, A., Van Cauwenberg, J., & Barnett, A. (2019). Neighbourhood built environment and older adults' physical activity. In A. P. Lane (Ed.), *Urban Environments for Healthy Ageing: A Global Perspective*: **Routledge**.

This chapter reviewed the existing evidence from both quantitative and qualitative studies on the relationship between the neighbourhood built environment and older adults' physical activity. Environmental features, classified into four categories (destination features, functional features, safety, and aesthetics and environmental quality) were examined in terms of their relationship with older adults' total physical activity and leisure- and transportation-related physical activity. The review showed that there was a strong correlation between the neighbourhood built environment and older adults' physical activity. To promote walking, especially for transport, a mix of neighbourhood amenities should be easily accessible. Leisure-related physical activity of older adults seemed to be supported when destinations were connected by a pedestrian-friendly street network. The authors also discussed the possibilities of future research such as a multi-country study.

Chui, E. (2012). Elderly learning in Chinese communities: China, Hong Kong, Taiwan and Singapore. In G. Boulton-Lewis & M. Tam (Eds.), *Active ageing, active learning: Issues and challenges* **(pp. 141–161). Springer**.

This chapter reviewed older adults' lifelong learning, i.e. 'elder learning' policies, in four predominantly Chinese areas: China, Taiwan, Hong Kong and Singapore. The author differentiated between formal, informal and non-formal modes of learning, and described elder learning as well as problems and prospects in the four places with respect to their socio-historical contexts. Despite differences in policies and programmes in the four places, the author concluded that elder learning was not only the responsibility of the state but should be promoted and cultivated from the ground as well. In view of lifelong learning's proven positive effects for active ageing, the author suggested for it to be adopted all over the world.

Clarke, P., & Nieuwenhuijsen, E. R. (2009). Environments for healthy ageing: A critical review. *Maturitas, 64*(1), 14–19.

In this paper, the current available health literature on environments for healthy ageing was reviewed with reference to the International Classification of Functioning Disability and Health framework. The authors emphasised the importance and role of supportive, barrier-free environments for older adults who were at a greater risk for developing disability and poor health. The surrounding environment characteristics were important elements that contributed to the independence, physical and mental health among older adults. While the majority of older adults preferred to age in place, these characteristics were not well understood. It was a dynamic process that involved interpersonal, social and environmental resource when an older adult managed between maintaining independence and declining health and function.

Dobner, S., Musterd, S., & Droogleever Fortuijn, J. (2014). 'Ageing in place': Experiences of older adults in Amsterdam and Portland. *GeoJournal, 81*(2), 197–209.

This qualitative study examined and compared older adults' experiences of ageing in place in two countries—Amsterdam, the Netherlands and Portland, USA. More specifically, the paper addressed social support and ties of older adults against the changing nature of welfare states. A total of 27 in-depth interviews with older adults aged between 65 and 94 years, and 13 interviews with key informants (e.g. members of neighbourhood and local organisations, churches, and governmental institutions) were conducted in two neighbourhoods in each city. The study found that the experiences of older adults in both cities were relatively similar despite the different political and social contexts. However, because of the different welfare systems in the two countries, there were more bottom-up and community-led informal networks in Portland, and more state-led support in Amsterdam.

Ellis, G., Hunter, R. F., Hino, A. A. F., Cleland, C. L., Ferguson, S., Murtagh, B., Anez, C. R. R., Melo, S., Tully, M., Kee, F., Sengupta, U., & Reis, R. (2018). Study protocol: Healthy urban living and ageing in place (HULAP): An international, mixed methods study examining the associations between physical activity, built and social environments for older adults the UK and Brazil. *BMC Public Health, 18*(1), 1135.

This study protocol outlined a multi-disciplinary study on the influence of the built environment on ageing in place in Belfast, UK, and Curitiba, Brazil. The study would employ and integrate quantitative (objective physical activity and sedentary behaviour measurement, GPS tracking, GIS, built environment audit and a survey) and qualitative (focus groups, interviews) methods, involving 300 participants in each city. This would be complemented with literature review, policy mapping, discourse analysis, development of walkability tools and neighbourhood audits to gain a deeper understanding of the social and built environment's influence on older adults' physical activity and urban governance that enabled healthy ageing. The expected outputs would be evidence, strong knowledge exchange and policy tools for increasing physical activity and wellbeing for ageing in place.

Fulvia, P., & Mina, S. (2018). Key characteristics of an age-friendly neighbourhood. *TeMA: Journal of Land Use, Mobility and Environment,* **117–132.**

Building upon the WHO Age-Friendly City framework, this paper reported on a literature review on accessible and socially inclusive qualities of the built environment. The first segment of the paper reviewed the literature on concepts of the urban environment, physical activity and walking to recognise factors that contributed to mobility. The following segment reviewed studies on urban quality of life and wellbeing in relation to the built environment. The paper concluded with a summary of associated effects of built environment factors and recommended for more mixed-method studies that targeted enacted mobility (assistive devices and public transportation) as possible areas for future studies.

Garin, N., Olaya, B., Miret, M., Ayuso-Mateos, J. L., Power, P., Bucciarelli, P., & Haro, J. M. (2014). Built environment and elderly population health: A comprehensive literature review. *Clinical Practice & Epidemiology in Mental Health, 10,* **103–115.**

This systematic review provided evidence on the relationship of built environment, ageing and health among community dwelling older adults from 2002 to 2013. Based on a review of 48 international studies, built environmental factors that might have an effect on health were identified including physical health (e.g. preventive care, functioning and disability, injuries, and other health-related conditions), mental health (e.g. depression, psychological distress), and life satisfaction (e.g. quality of life, well-being). The authors acknowledged the methodological limitations of current evidence and recommended further longitudinal studies with consideration of reliability and comparability to gain a better understanding of these relationships, and in turn might lead to public policy implications.

Handler, S. (2018). Alternative age-friendly initiatives: Redefining age-friendly design. In S. Handler, T. Buffel, & C. Phillipson (Eds.), *Age-friendly Cities and Communities* **(1 ed., pp. 211–230): Bristol University Press.**

This chapter posited that the arguably rigid and domain-structured tenets of age-friendliness had stunted creative proactivity for age-friendly policies, and there was a need to encourage critical reflection beyond the disciplinary boundaries of design practice. In particular, socially engaged designer should be encouraged to use innovative design tools (e.g. filmmaking, storytelling, minimalistic interventions) that promoted critical reflection on age-friendly policies. The concept of age-friendliness was best viewed not just as a rights-based discourse of objective needs but also emotional geographies such as subjective connections, longing and desires that formed an integral element of older people's rights to the city.

Kalache, A. (2016). Active ageing and age-friendly cities—A personal account. In T. Moulaert & S. Garon (Eds.), *Age-Friendly Cities and Communities in International Comparison: Political Lessons, Scientific Avenues, and Democratic Issues* **(Vol. 14, pp. 65–77). Springer.**

This chapter elaborated the history, notions and international development of 'active ageing' and the WHO Age-Friendly Cities using a self-interview with Dr. Alexandre Kalache who previously directed the WHO global ageing programme. The conversation could be separated into three parts including (1) biographical background describing Dr Kalache's professional pathway towards ageing and global public health; (2) history of how 'active ageing' came into being within the World Health Organization (WHO) and was developed beyond 'healthy ageing' and into an international strategy on ageing; (3) transition of Age-Friendly Cities as a feasible conception of active ageing. The author highlighted the role of the WHO to promote 'active ageing' and Age-Friendly Cities as well as academic research as a foundation for the understanding of these notions and the main challenges.

Kano, M., Rosenberg, P. E., & Dalton, S. D. (2018). A global pilot study of age-friendly city indicators. *Social Indicators Research, 138*(3), 1205–1227.

This study was an evaluative assessment of the measurability of core indicators developed under the WHO Age-Friendly City initiative. Some challenges faced in measurement were the lack of data on social indicators of community quality of life and community-based health. The paper argued for a need to improve the social metrics for age-friendly initiatives. The main lessons from the evaluative study were the need to keep indicators flexible to accommodate data variance and to design core indicators that were locally adaptable to facilitate more meaningful equity analysis.

Kerr, J., Rosenberg, D., & Frank, L. (2012). The role of the built environment in healthy aging: Community design, physical activity, and health among older adults. *Journal of Planning Literature, 27*(1), 43–60.

This paper reviewed the current literature to gain a deeper understanding of the role of community design on health, specifically physical activity and travel behaviour, in later life. Existing studies on the association of built environment, walking and health outcomes among older adults demonstrated age-friendly community design. For example, walkable communities and accessible key destinations were protective factors of chronic disease and encouraged active and independent life of older adults. The authors emphasised that destinations, safely accessible on foot or by transit, should be considered from planning and design perspectives.

Lee, C., & Zhong, S. (2019). Age-friendly communities: Community environments to promote active ageing in place. In A. P. Lane (Ed.), *Urban Environments for Healthy Ageing: A Global Perspective.* **London: Routledge.**

This chapter examined the neighbourhood built and social environments of age-friendly communities that supported healthy and active ageing. Literature and existing data such as best practice case studies and international examples of age-friendly activities were reviewed. The focus was on those elements of the built environment that were associated with physical activity among older adults such as housing, land use mix and density, street networks, parks and open spaces, public transport, and neighbourhood attractiveness. The authors also drew attention to the importance of intergenerational communities and interactions for healthy and active ageing in place.

Loo, B. P. Y., Lam, W. W. Y., Mahendran, R., & Katagiri, K. (2017). How is the neighborhood environment related to the health of seniors living in Hong Kong, Singapore, and Tokyo? Some Insights for Promoting Aging in Place. *Annals of the American Association of Geographers, 107*(4), 812–828.

This study assessed the neighbourhood environmental factors that were associated with physical and mental health of older adults in 3 cities in 3 countries—Hong Kong, Singapore and Tokyo. Observational and questionnaire surveys were implemented with 687 older people aged 65 and above in the 3 cities. Multilevel analysis revealed that 17.53 and 8.24% variability of general neighbourhood factors explained physical and mental health respectively while the rest of the variability depended on individual determinants including personal and individual-based neighbourhood factors. Biological factors (e.g. age, gender) were found to be not as important as being in the normal range of weight (BMI) and using walking aid properly (which encouraged older adults aged 85 years and above to be more active in the community). The findings indicated that the association of local neighbourhood factors with health was independent of the city in which the older person was living.

Mahmood, A., Chaudhury, H., Michael, Y. L., Campo, M., Hay, K., & Sarte, A. (2012). A photovoice documentation of the role of neighborhood physical and social environments in older adults' physical activity in two metropolitan areas in North America. *Social Science & Medicine, 74*(8), 1180–1192.

This study examined the physical and social environmental determinants, which had an impact on physical activity of older adults in two age-friendly cities, metropolitan Vancouver, Canada and Greater Portland, Oregon, USA. Successive approximation, a deductive analysis method, was applied on data from participatory photovoice with 66 community dwelling older adults aged 65 years and older in eight neighbourhoods. Seven themes related to physical and social aspects of the neighbourhood were explored including getting there, diversity of destinations, community-based programmes, physical-dominant features (i.e. being safe and feeling secure, comfort in movement), and social-driven factors (i.e. peer support, intergenerational/volunteer activities). Facilitators and obstacles for both daily activity and planned physical activities were identified among these environmental determinants. Findings indicated the importance of physical and social environmental features in encouraging healthy habits among older adults.

Moulaert, T., & Garon, S. (2015). Researchers behind policy development: Comparing 'age-friendly cities' models in Quebec and Wallonia. *Journal of Social Work Practice, 29*(1), 23–35.

This paper aimed to address the role and place of researchers in the development of Age-friendly cities and Environments by comparing two age-friendly cities/regions in two countries, Walloon region, Belgium and Quebec province, Canada. Adopting reflexive method and drawing on public sociology, empirical data was collected from the respective cities' age-friendly cities projects. Findings indicated the absence of research and the researcher in Walloon while a more developed practice of public sociology and involvement in research was demonstrated in Quebec including continuous knowledge transfer as an ongoing public open dialogue. The authors recommended 'knowledge in action' and 'action in knowledge' as essential steps of a dialogic communication among all stakeholders (i.e. older people, city officers, local and central policymakers, and researchers) in the making of age-friendly environments.

Moulaert, T., Boudiny, K., & Paris, M. (2016). Active and healthy ageing: Blended models and common challenges in supporting age-friendly cities and communities. In T. Moulaert & S. Garon (Eds.), *Age-friendly cities and communities in international comparison: Political lessons, scientific avenues, and democratic issues* **(Vol. 14, pp. 277–304). Springer**.

This chapter emphasised the significance of the origins of active and healthy ageing behind the WHO Age-Friendly Cities and Communities. It argued for an integrative framework with empowerment as the central element in the implementation of active and healthy ageing, and proceeded to elaborate from two applicable aspects—multilevel perspective and an insider's view, to integrate from theoretical grounds into practice. The authors acknowledged the potential prospect of applying the integrative framework, such as placing empowerment as a key element, to address the diverse ageing profiles in age-friendly cities.

Neville, S., Wright-St Clair, V., Montayre, J., Adams, J., & Larmer, P. (2018). Promoting age-friendly communities: An integrative review of inclusion for older immigrants. *Journal of Cross-Cultural Gerontology*, *33*(4), 427–440.

This paper reviewed the literature—empirical studies focused on the inclusiveness of communities for older immigrants. Multiple factors were identified in connection with perceived inclusiveness: (1) understanding of the local language, (2) reception of financial and non-financial support, and (3) having the ability to participate in community activities. Findings indicated that social inclusion was positively associated with mental health. The paper concluded with recommendations for policymakers to consider the contextual experiences of older immigrants in order to develop more age-friendly communities.

Phillipson, C. (2011). Developing age-friendly communities: New approaches to growing old in urban environments. In R. A. Settersten & J. L. Angel (Eds.), *Handbook of Sociology of Aging* (pp. 279–293). Springer.

Through a socio-economic approach, this paper argued that research and policy agenda would need to change in three ways to ensure a higher quality of life for older citizens in age-friendly cities. First, ageing research should increasingly consider the impacts of global and economic forces on urban environments. This was important as these forces played a significant role in shaping the physical and social landscape of cities. Second, research should increasingly aim to study the effects of urban growth on the experiences of minority (e.g. lower income, singles or migrant) groups to understand the challenges faced by different groups. Third, it was critical that social and physical infrastructure worked in tandem to nurture age-friendly values that promoted harmonious city living.

Plouffe, L., & Kalache, A. (2010). Towards global age-friendly cities: Determining urban features that promote active aging. *Journal of Urban Health, 87*(5), 733–739.

This paper presented the development of the WHO Global Age-Friendly Cities Guide and Checklist of Essential Features of Age-Friendly Cities to improve and assist cities towards active aging. Eight domains of age-friendly cities in urban life were identified by the WHO and collaborating 35 cities, involving 158 semi-structured focus groups with 1485 older adults (aged 60 years and above), 250 caregivers, and 515 service providers. The 8 domains were: outdoor spaces and buildings; transportation; housing; social participation; respect and social inclusion; civic participation and employment; communication and information; and community support and health services. Content analysis of age-friendly features, barriers, and suggestions for improvement under each topic area showed no major differences between developed and developing countries though more positive and age-friendly features were presented in the cities of developed countries while physical accessibility, proximity, security, affordability, and inclusiveness were explored as important features in all study cities.

Plouffe, L. A., & Kalache, A. (2011). Making communities age friendly: State and municipal initiatives in Canada and other countries. *Gaceta Sanitaria, 25*, 131–137.

This paper reported and compared the development of age-friendly community initiatives in Canada, Spain, Brazil and Australia. Descriptive analysis was conducted on working reports, historical documents, literature publication and communication with stakeholders of these initiatives. Strategical engagement and policy action were found as common strategies of age-friendly initiatives across the study countries while knowledge development and exchange were specifically explored in Canada.

Although these initiatives were in the early stage of implementation with limited evaluation, the authors endorsed the improvements in community-level policies, practices and design for active ageing, and highlighted the importance of older adults' involvements in municipal and state planning and policy.

Plouffe, L., Kalache, A., & Voelcker, I. (2016). A critical review of the WHO age-friendly cities methodology and its implementation. In T. Moulaert & S. Garon (Eds.), *Age-friendly cities and communities in international comparison: Political lessons, scientific avenues, and democratic issues* **(Vol. 14, pp. 19–36). Springer.**

This book chapter reviewed the implementation of the Vancouver Protocol and WHO checklist of age-friendly city features in a variety of municipalities, regions, states and countries in North and South America, Europe and Australia in order to determine the lessons learned. Literature review using published reports, online materials and grey literature in English, French, Portuguese and Spanish as well as first-hand accounts from people involved in their implementation were used in the evaluation. Findings indicated that while the WHO age-friendly domains were generally relevant, some features were less relevant for older people in less developed countries, e.g. the availability of housing options and community support and health services. Additions and refinements were made in various locations, e.g. Quebec and Lyon had distinguished civic participation from other forms of participation to emphasise the engagement of older people as volunteers to formal and informal community support. The conclusion was not to focus on which domains were the right ones but to note the interdependence among the features and the need to regard them as a dynamic whole and importantly, to promote the active participation of older people in the process.

Redondo, N., & Gascón, S. (2016). The implementation of age-friendly cities in three districts of Argentina. In T. Moulaert & S. Garon (Eds.), *Age-friendly cities and communities in international comparison: Political lessons, scientific avenues, and democratic issues* **(Vol. 14, pp. 153–170). Springer.**

This chapter analysed and compared the implementation and research on the Age-Friendly Cities programme in three cities (La Plata, Resistencia and Lezama) in Argentina. Using the 8 WHO domains in the Vancouver Protocol, 207 older adults aged 60 years and above, informal caregivers, state providers, private providers, and NGO providers were interviewed. Positive and negative aspects of life for older adults were identified. Findings showed that city size, level of human, economic and social development, and urban design were variables that determined older people's differing quality of life and access to full social inclusion.

Rémillard-Boilard, S. (2018). The development of age-friendly cities and communities. In T. Buffel, S. Handler, & C. Phillipson (Eds.), *Age-friendly cities and communities* **(1 ed., pp. 13–32): Bristol University Press.**

In reference to the WHO Age-Friendly Cities and Communities model, the chapter provided an assessment of the key challenges and successes of the model. Findings indicated that among others, the sustainability of age-friendly policies was threatened due to budget constraints, and minority ethnic groups might become increasingly marginalised due to a lack of resources to consider their needs. The author argued that addressing social exclusion in the face of urbanisation pressures would be crucial to the creation of inclusive age-friendly city projects.

Scharlach, A. E. (2016). Age-friendly cities: For whom? By whom? For what purpose?. In T. Moulaert & S. Garon (Eds.), *Age-friendly cities and communities in international comparison: Political lessons, scientific avenues, and democratic issues* **(Vol. 14, pp. 305–329). Springer**.

This chapter explored the meaning and conceptualisation of the age-friendly cities framework. The idea of age-friendliness was deconstructed into liveability, elder-friendliness and ageing- friendliness with emphasis on the communal and transactional aspects of age-friendliness. A critical review was conducted of age-friendly characteristics, their interrelated nature, the needs of different groups of older adults and stakeholders, etc. The conclusion was that while there had been previous studies on age-friendly characteristics, there was far less research on age-friendly city implementation. Implementation was, however, not uniform, and three types of age-friendly city initiatives were identified for further investigation: community-wide planning, inter-organisational collaboration, and community development.

Steels, S. (2015). Key characteristics of age-friendly cities and communities: A review. *Cities, 47*, **45–52**.

This paper examined the current age-friendly interventions that had arisen from the WHO Age-Friendly Cities and Communities framework. Using literature review methodology, 64 papers were reviewed. Successful interventions were characterised as those that have: (1) multi-stakeholder participation, (2) inclusive processes that allowed for co-planning or co-design, and (3) policies that considered both physical and social environments. In the face of scarce resources, planners were required to justify resources spent on interventions while existing ways of evaluation were too context specific and lack generalisability. The author argued for a framework or set of indicators to assess the success or failure of age-friendly interventions. The aspiration was to develop an international adaptable framework tool that would further assist age-friendly policymaking.

Sun, Y., Chao, T. Y., Woo, J., & Au, D. W. H. (2017). An institutional perspective of "Glocalization" in two Asian tigers: The "Structure − Agent − Strategy" of building an age-friendly city. *Habitat International, 59*, **101–109**.

This paper examined age-friendly initiatives in the Asian context through a comparison of Hong Kong and Taiwan—Chiayi City's glocalization of the WHO Age-Friendly Cities approach. Applying comparative institutional analysis, glocalization was examined in terms of how Hong Kong and Taiwan applied a globally oriented social policy locally, how different local actors, structures and strategies were deployed. Empirical data was collected through semi-structured interviews with those who were involved in the age-friendly projects. Results revealed three main findings: (1) local policies were crucial in avoiding fragmentation and achieving integration of resources, (2) in the case of Taiwan, local action was effected through the state whereas in Hong Kong it was through a grassroots approach involving non-governmental organisations, universities, etc., and (3) academia promoted age-friendly action in both cities.

Ulfarsson, G. F. and Kim, S. (2019). Ageing and transportation. In A. P. Lane (Ed.), *Urban environments for healthy ageing: A global perspective.* **London: Routledge**.

This chapter addressed transportation issues, mobility behaviour, traffic safety and quality of life of older adults in various countries including the US, UK, Canada, Australia, South Korea, among others. The aim was to illuminate potential future issues related to the mobility of older adults, which could inform urban planners and policymakers when developing future transportation options. The authors highlighted the needs of a vulnerable group of older adults—the captive car users who relied on cars as their main transport mode due to living in remote areas and were at risk of losing their mobility once they were not able to drive anymore. Alternative age-friendly transportation options should be provided for such people.

Warth, L. (2016). The WHO Global Network of Age-Friendly Cities and Communities: Origins, developments and challenges. In T. Moulaert & S. Garon (Eds.), *Age-friendly cities and communities in international comparison: Political lessons, scientific avenues, and democratic issues* **(Vol. 14, pp. 37–46). Springer**.

This chapter was an interview with Lisa Warth, a staff member of the World Health Organization (WHO) that had coordinated the WHO Global Network of Age-friendly Cities and Communities since 2012. The interviewee spoke about the origins of the Network, which was based on the United Nations Principles for Older Persons, social determinants of health, Active Ageing Policy Framework and Age-friendly Cities and Communities Guide. The overall approach of the WHO age-friendly cities project consisted of political commitment, participatory approach, coordination and collaboration across sectors, evidence-informed and results-oriented planning, and monitoring and evaluation of progress over time. The interviewee also touched upon the mission of the Network, listed its achievements and discussed the challenges for the future.